Apress Pocket Guides

Apress Pocket Guides present concise summaries of cutting-edge developments and working practices throughout the tech industry. Shorter in length, books in this series aims to deliver quick-to-read guides that are easy to absorb, perfect for the time-poor professional.

This series covers the full spectrum of topics relevant to the modern industry, from security, AI, machine learning, cloud computing, web development, product design, to programming techniques and business topics too.

Typical topics might include:

- A concise guide to a particular topic, method, function or framework
- Professional best practices and industry trends
- A snapshot of a hot or emerging topic
- Industry case studies
- Concise presentations of core concepts suited for students and those interested in entering the tech industry
- Short reference guides outlining 'need-to-know' concepts and practices.

More information about this series at `https://link.springer.com/bookseries/17385`.

Building a Virtual Assistant for Raspberry Pi

The Practical Guide for Constructing a Voice-Controlled Virtual Assistant

Second Edition

Harshil Agrawal
Tanay Pant

Apress®

Building a Virtual Assistant for Raspberry Pi: The Practical Guide for Constructing a Voice-Controlled Virtual Assistant, Second Edition

Harshil Agrawal
Berlin, Germany

Tanay Pant
Berlin, Berlin, Germany

ISBN-13 (pbk): 979-8-8688-1759-5
https://doi.org/10.1007/979-8-8688-1760-1

ISBN-13 (electronic): 979-8-8688-1760-1

Copyright © 2025 by Harshil Agrawal and Tanay Pant

This work is subject to copyright. All rights are reserved by the Publisher, whether the whole or part of the material is concerned, specifically the rights of translation, reprinting, reuse of illustrations, recitation, broadcasting, reproduction on microfilms or in any other physical way, and transmission or information storage and retrieval, electronic adaptation, computer software, or by similar or dissimilar methodology now known or hereafter developed.

Trademarked names, logos, and images may appear in this book. Rather than use a trademark symbol with every occurrence of a trademarked name, logo, or image we use the names, logos, and images only in an editorial fashion and to the benefit of the trademark owner, with no intention of infringement of the trademark.

The use in this publication of trade names, trademarks, service marks, and similar terms, even if they are not identified as such, is not to be taken as an expression of opinion as to whether or not they are subject to proprietary rights.

While the advice and information in this book are believed to be true and accurate at the date of publication, neither the authors nor the editors nor the publisher can accept any legal responsibility for any errors or omissions that may be made. The publisher makes no warranty, express or implied, with respect to the material contained herein.

    Managing Director, Apress Media LLC: Welmoed Spahr
    Acquisitions Editor: Miriam Haidara
    Editorial Assistant: Jessica Vakili

Cover designed by eStudioCalamar

Distributed to the book trade worldwide by Springer Science+Business Media New York, 1 New York Plaza, New York, NY 10004. Phone 1-800-SPRINGER, fax (201) 348-4505, e-mail orders-ny@springer-sbm.com, or visit www.springeronline.com. Apress Media, LLC is a Delaware LLC and the sole member (owner) is Springer Science + Business Media Finance Inc (SSBM Finance Inc). SSBM Finance Inc is a **Delaware** corporation.

For information on translations, please e-mail booktranslations@springernature.com; for reprint, paperback, or audio rights, please e-mail bookpermissions@springernature.com.

Apress titles may be purchased in bulk for academic, corporate, or promotional use. eBook versions and licenses are also available for most titles. For more information, reference our Print and eBook Bulk Sales web page at http://www.apress.com/bulk-sales.

Any source code or other supplementary material referenced by the author in this book is available to readers on GitHub. For more detailed information, please visit https://www.apress.com/gp/services/source-code.

If disposing of this product, please recycle the paper

*To my dad, who has always supported me,
even when he is not around.*

—Harshil Agrawal

*To my loving wife, Isabel, and my family, who have always
inspired and supported me.*

—Tanay Pant

Table of Contents

About the Authors xi

About the Technical Reviewer xiii

Acknowledgments xv

Introduction xvii

Chapter 1: Introduction to Virtual Assistants 1

Commercial Virtual Assistants 2

Raspberry Pi 3

How a Virtual Assistant Works 3

    Speech Recognition Engine 4

    Logic Engine 4

    Voice Engine 5

Setting Up Your Development Environment 5

    Python 3.x 5

    Python Package Index (PyPI) 6

    Version Control System (Git) 6

Designing Melissa 7

    Save Project on GitHub 8

    Virtual Environment 9

    Vosk 10

    PyAudio 10

TABLE OF CONTENTS

Learning Methodology 11

Get the Code 11

Summary 12

Chapter 2: Understanding and Building an Application with STT and TTS 15

Speech-to-Text Engine 15

    Vosk 16

    Melissa's Inception 20

Text-to-Speech Engine 21

    macOS 21

    Linux 22

    Implementing the TTS Engine 23

    Adding a Personal Touch 24

Get the Code 25

Summary 26

Chapter 3: Making Melissa Intelligent 27

Introduction to Large Language Models 27

    Using an LLM 28

Installing Ollama 30

    Use the Local Server 32

Implement the Logic Engine for Melissa 33

    Interact with Melissa 36

Get the Code 38

Summary 38

TABLE OF CONTENTS

Chapter 4: Introducing Plugin Architecture 41
What Is a Plugin? 41
Create Your First Plugin 42
Function Calling in LLMs 44
Get the Code 51
Summary 52

Chapter 5: Get Top Tech Stories from Hacker News 53
What Is an API? 53
Creating the Hacker News Plugin 54
Integrating the Plugin with Melissa 58
Get the Code 62
Summary 62

Chapter 6: Mellisa, Tell Me the Weather! 63
Get OpenWeather API Key 63
Create the Weather Plugin 65
Melissa, What Is the Temperature? 67
Securely Storing the API Key 72
Get the Code 79
Summary 80

Chapter 7: Saving Notes with Melissa 81
Understanding Local Data Storage 81
    Why SQLite? 82
Creating the Notes Plugin 83
    Save Note 84
    Get Note 85
    List Notes 86

ix

TABLE OF CONTENTS

    Update Note 87
    Delete Note 88
    Search Note 88

Integrating the Notes Plugin with Melissa 90

Testing the Notes Plugin 99

Get the Code 102

Summary 102

Chapter 8: What More Can You Build? 103

Remind Yourself to Call Dad 103

What's Next on My Schedule? 104

Control the Lights with Voice 104

When Is the Next Bus? 105

Share Your Thoughts with the World! 105

Play My Favorite Song 107

Summary 108

Chapter 9: Integrating the Software with Raspberry Pi, and Next Steps 111

Setting Up a Raspberry Pi 5 112

    Installing the Operating System (Raspberry Pi OS) 113
    Setting Up VNC (Remote Desktop Access) 119
    Adding New Components to the Raspberry Pi 5 123

Setting Up Melissa 124

Making Melissa Better Each Day! 126

Where Do I Use Melissa? 128

Summary 129

Index 131

About the Authors

Harshil Agrawal is an educator, public speaker, builder, and developer relations professional. He is known for educating the developer community through blog articles and videos. He enjoys building games and applications for everyday use.

Tanay Pant is an author, speaker, educator, and developer relations expert. He is best known for his work with early-stage tech startups and various books on computer science. He has been listed on Firefox's about:credits page for his contributions to the Mozilla Foundation's different open-source projects.

About the Technical Reviewer

Pratik Parmar is an entrepreneur, AI nerd, and developer relations expert who is passionate about empowering the developer community through knowledge sharing, talks, and tutorials. When he's not staring at his screen, he loves trekking and spending time with his horses.

Acknowledgments

This book would not exist without the extraordinary support of many people.

First, my deepest gratitude to Tanay Pant, who gave me an opportunity to write this book. His continued feedback and gentle guidance helped shape this manuscript into something far better than I could have imagined.

My friends, who kept me motivated throughout the process and believed in the ideas I shared with them. To Carmen Huidobro, who was supportive and helpful in the early works of this book.

Finally, to my family, who helped me become the man I am today. They instilled in me the curiosity of learning new things and taught me that failure is a part of life. This book is the result of curiosity and perseverance of learning and building something new!

—Harshil Agrawal

I would like to thank my co-author, Harshil. Collaborating with him is always a great experience, and I have learned a lot while working on this project with him. My heartfelt gratitude to all those who've built the fantastic tools we've used in this book.

—Tanay Pant

Introduction

This book covers all my learnings of building an "intelligent" personal assistant. I started building a personal assistant reading the first edition of this book. It was a great start, and I wanted to take it a step further.

In this book, you will learn about various technologies, libraries, and services. You will start with building a Python application that takes voice as an input, processes the input, and returns the output as audio, making your assistant purely voice based!

To make your assistant smart, you will also use artificial intelligence (AI). You will learn about large language models (LLMs), which are changing our world. You will install and run an LLM model locally on your computer and use it to make your assistant intelligent.

While LLMs are good at answering general questions, they can't perform tasks on your behalf. For that, you will learn and implement function calling and follow a plugin architecture. This will allow your assistant to follow your instructions and perform tasks on your behalf.

Finally, you will install and configure the application to your Raspberry Pi.

CHAPTER 1

Introduction to Virtual Assistants

Building a Virtual Assistant for Raspberry Pi is a short step-by-step guide that will teach you how to build a personal virtual assistant. You will learn about the components involved, get a better understanding of large language models (LLMs), and get practical knowledge by implementing various components.

This chapter gives a detailed overview of what smart virtual assistants are, common assistants in the market, what qualities a smart virtual assistant should possess, and the basic workflow and design for building a scalable smart virtual assistant. You will also learn about the various tools required to build Melissa (your own smart virtual assistant) in upcoming chapters and the methodology you follow in this book.

The advent of smart virtual assistants has been an important event in the history of computing. Virtual assistants are useful for helping the users of a computer system automate tasks and accomplish tasks with minimum human interaction with a machine. The interaction that takes place between a user and a virtual assistant seems natural; the user communicates using their voice, and the software responds in the same way. A smart virtual assistant takes this a step forward. With the advancement in AI and increasing development of LLMs, virtual assistants are smarter.

CHAPTER 1   INTRODUCTION TO VIRTUAL ASSISTANTS

If you have seen the movie *Iron Man*, you can perhaps imagine having a virtual assistant like Tony Stark's Jarvis. Does that idea excite you? The movie has always been an inspiration for us to build our own smart virtual assistant, Melissa. Such an assistant can serve in the Internet of Things as well as run a voice-controlled coffee machine or a voice-controlled drone. Let's take a look at some of the virtual assistants that are available in the market.

Commercial Virtual Assistants

Virtual assistants are useful for carrying out tasks such as saving notes, telling you the weather, playing music, retrieving information, and doing much more. The following are some virtual assistants that are already available in the market:

- **Google Assistant**: Developed by Google for Android and iOS mobile operating systems. It also runs on computer systems with the Google Chrome web browser. The best thing about this software is its voice recognition ability.

- **Co-pilot**: Developed by Microsoft and runs on Windows for desktop. It also runs on both Android and iOS. Co-pilot doesn't entirely rely on voice commands: you can send commands by typing.

- **Siri**: Developed by Apple and runs only on iOS, macOS, watchOS, and tvOS. Siri is an advanced personal assistant with lots of features and capabilities.

CHAPTER 1   INTRODUCTION TO VIRTUAL ASSISTANTS

These are sophisticated software applications that are proprietary in nature.

So, you can't run them on a Raspberry Pi. But you can build one yourself!

Raspberry Pi

The software you are going to create should be able to run with limited resources. Even though you are developing Melissa from laptop/desktop systems, you will eventually run this on a Raspberry Pi.

The Raspberry Pi is a credit-card-sized, single-board computer developed by the Raspberry Pi Foundation for the purpose of promoting computer literacy among students. The Raspberry Pi has been used by enthusiasts to develop interesting projects of varying genres and by the industry to serve the Internet of Things. In this book, you will build a voice-controlled virtual assistant named Melissa to control this little computer with your voice.

This project uses a Raspberry Pi 5. You can find information on where to purchase it at https://www.raspberrypi.com/products/raspberry-pi-5/. Do not worry if you don't currently have a Raspberry Pi; you will carry out the complete development of Melissa on a *nix-based system.

How a Virtual Assistant Works

Let's discuss how Melissa works. Theoretically, such software primarily consists of three components: the speech recognition engine, the logic-handling engine, and the voice engine (see Figure 1-1).

CHAPTER 1   INTRODUCTION TO VIRTUAL ASSISTANTS

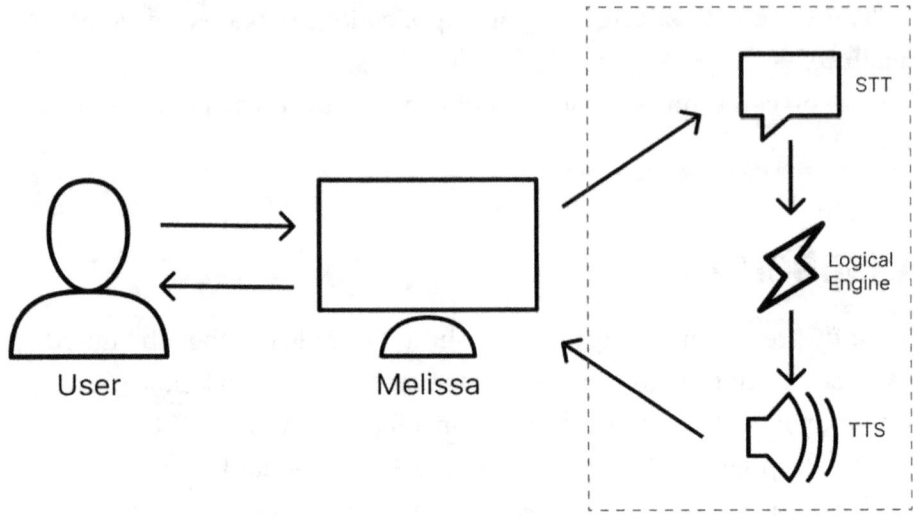

Figure 1-1. *Virtual assistant workflow*

Speech Recognition Engine

As the name suggests, the speech recognition engine converts the user's voice input into a text string that can be processed by the logic engine. This involves recording the user's voice, capturing the words from the recording (canceling any noise and fixing distortion in the process), and then using natural language processing (NLP) to convert the recording to a text string.

Logic Engine

Melissa's logic engine is the software component that receives the text string from the speech recognition engine and handles the input by processing it and passing the output to the voice engine. The logic engine can be considered Melissa's brain; it handles user queries via an LLM. It decides what the output should be in response to specific inputs. You build Melissa's logic engine throughout the book, improving it and adding new functionalities and features as you go. You learn more about the logic engine and how to implement it in Chapter 3.

Voice Engine

This component receives the output from Melissa's logic engine and converts the string to speech to complete the interaction with the user. Text-to-speech (TTS) is crucial for making Melissa more humane, compared to giving confirmation via text.

This three-component system removes any physical interaction between the user and the machine; the users can interact with their system the same way they interact with other human beings. You learn more about the speech-to-text (STT) and TTS engines and how to implement them in Chapter 2.

From a high-level view, these are the three basic components that make up Melissa. This book shows you how to do all the necessary programming to develop them and put them together.

Setting Up Your Development Environment

This is a crucial section that is the foundation of the book's later chapters. You need a computer running a *nix-based operating systems such as Linux or macOS. We are using a MacBook running macOS for the purpose of illustration.

Python 3.x

You will write Melissa's code in the Python programming language. So, you need to have the Python interpreter installed to run the Python code files. *nix systems generally have Python preinstalled. You can check whether you have Python installed by running the following command in the terminal of your operating system:

```
$ python3 --version
```

This command returns the version of the Python installed on your system. In our case, it gives the following output:

```
Python 3.12.4
```

This should also work on other versions of Python 3.

If you don't have Python installed, you can follow the official Python Download page (https://www.python.org/downloads/).

Python Package Index (PyPI)

You need pip to install the third-party modules that are required for various software operations. You use these third-party modules, so you do not have to reinvent the wheels of assorted basic software processes.

You can check whether pip is installed on your system by issuing the following command:

```
$ pip --version
```

In my case, it gives this output:

```
pip 24.0 from /usr/local/lib/python3.12/site-packages/pip (python 3.12)
```

If you do not have pip installed, you can install it by following the guide at https://pip.pypa.io/en/stable/installing/.

Version Control System (Git)

You use Git for version control of your software as you work on it to avoid losing work due to hardware failure or system administrator mistakes. You can use GitHub to upload your Git repository to an online server. You can check whether you have Git installed on your system by issuing the following command:

```
$ git --version
```

This command gives me the following output:

```
git version 2.45.2
```

If you do not have Git installed, you can install it using the instructions at http://git-scm.com/downloads.

Designing Melissa

You will follow the don't repeat yourself (DRY) and keep it simple, stupid (KISS) principles and use modular code to design Melissa. Doing so helps maintain your code properly and makes it easier to scale the code in the future when you want to add cool features to your existing codebase. So, let's first design the structure of your code directories:

```
.gitignore
src/
    ...
    __init__.py
    main.py
    tts.py
    stt.py
    logic.py
profile.json
requirements.txt
```

In this directory structure, ... denotes that files will be added here in the future as you go through the chapters in this book. The folder containing `__init__.py` file is a Python package. The `main.py` file will be the entry point of the program where you will make use of the STT, logic, and TTS engines. The `requirements.txt` file will be used for keeping tabs on the third-party Python modules you use in this project.

CHAPTER 1   INTRODUCTION TO VIRTUAL ASSISTANTS

The `profile.json` file will store information such as your name and city where you live, in JSON format. The `profile.json` file is crucial for executing the main.py file since it will provide Melissa some information about you. It helps make the assistant more personalized. You can update the file to add details like your favorite music genre, and Melissa will use this information when handling your request, e.g., playing your music!

Currently the contents of `profile.json.default` are as follows:

```
{
"name": "John Doe",
"city_name": "New York"
}
```

The contents of the `.gitignore` file are as follows:

```
venv/
__pycache__/
*.pyc
profile.json
```

Now that you know the high-level directory structure of the project, you can go ahead and create the skeleton structure. This structure will help you keep the codebase clean and properly organized as you move through the book and work on building new features.

Save Project on GitHub

After you install the required software and create the project, the next step is to initialize Git for the project. You can use the following command to do that:

```
$ git init
```

You can use Git to version control the project, as well as push the code on GitHub, GitLab, or BitBucket. This will let you save a copy of the project on the cloud and install and use it on any machine you want. You can also share the repository with your friends and help them build a virtual assistant for themselves!

In this book, we will use GitHub. But if you use GitLab or BitBucket, the commands will still be the same.

To add the project to GitHub, you need to create an empty GitHub repository. You can create a new GitHub repository on https://github.new. After you create the repository, copy the URL of the repository. You will use it in the next step.

Use the following commands to commit the changes you have made so far, configure the remote (GitHub) repository, and push the code to GitHub. Make sure to replace <GITHUB_REPO_URL> with the actual URL you copied earlier.

```
$ git add .
$ git commit -m "initialized project"
$ git remote add origin <GITHUB_REPO_URL>
$ git push origin main
```

On success, you should see the project code available in your GitHub repository.

Virtual Environment

You created the project directory and initialized it with Git. In the next sections, you will learn how to execute Python code and install packages using pip.

To run the Python code, you will create and use a virtual environment. A virtual environment allows your Python project to run in isolation. All the packages required for your project gets installed in this virtual isolated environment. This isolation helps you use different versions of the same Python libraries in different projects, which prevents conflicts with globally installed packages.

To create a virtual environment for your project, you can use the following command:

```
$ python3 -m venv venv
```

To activate this virtual environment, you can use the following command:

```
source venv/bin/activate
```

Vosk

Vosk (`https://alphacephei.com/vosk/`) is a lightweight STT engine that runs entirely offline. It is well-suited for low-resource devices like the Raspberry Pi, making it an excellent choice for projects requiring voice recognition without an internet connection. Since Vosk runs without the internet, your conversations are private and are not shared with anyone on the internet.

PyAudio

PyAudio allows Python to record and play audio on a variety of platforms, which is exactly what you need for your STT engine. You can find the instructions for installing PyAudio at `http://people.csail.mit.edu/hubert/pyaudio/`.

You also need a microphone via which you can speak to your computer (and perform voice recording) and speakers to hear the output. Most modern laptops have these installed by default. For a Raspberry Pi, you need an external microphone and speakers/earphones.

Learning Methodology

This section describes the methodology you use throughout the book: understanding concepts, learning by prototyping, and then developing production-quality code to integrate into the skeleton structure you just developed (see Figure 1-2).

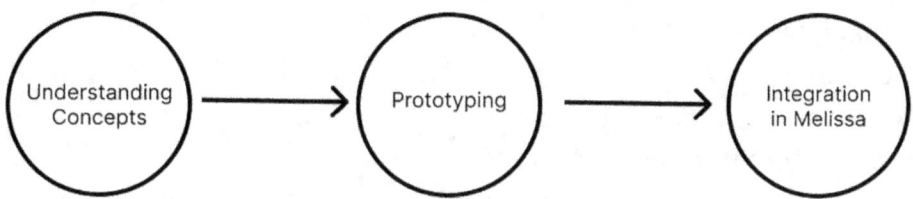

Figure 1-2. *Learning methodology*

First you explore the theoretical concepts as well as understand the core principles that will enhance your creativity and help you see different ways to implement features. This part may seem boring to some people, but do not skip these bits.

Next, you implement your acquired knowledge in Python code and play around with it to convert your knowledge into skills. Prototyping will help you to understand the functioning of individual components without the danger of messing up the main codebase. Finally, you edit and refactor the code to create good-quality code that can be integrated with the main codebase to enhance Melissa's capabilities.

Get the Code

If you have skimmed through the chapter and are interested in just the code, you can find the code in this GitHub repository: https://github.com/Melissa-AI/melissa-v2.

Clone the repository and start a new virtual environment. Install the packages and update the `profile.json` file with your details. Note that if you are adding personal information in the `profile.json` file, make sure you don't commit it.

```
// clone the repo
$ git clone https://github.com/Melissa-AI/melissa-v2.git
// navigate into the repo
$ cd melissa-v2
// switch to chapter 1 branch
$ git checkout chapter-1
// create the virtual environment
$ python3 -m venv venv
// start the virtual environment
$ source venv/bin/activate
// install packages
$ pip install -r requirements.txt
```

On macOS, if you run into errors related to `pyaudio` or `portaudio`, you may need to install `portaudio`. To install `portaudio`, use the following command:

```
$ brew install portaudio
```

Summary

In this chapter, you learned about what virtual assistants are. You also saw various virtual assistants that exist in the commercial market, the features a virtual assistant should possess, and the workflow of a voice-controlled virtual assistant. You designed Melissa's codebase structure and were introduced to the methodology that this book follows to create an effective learning workflow.

In the next chapter, you will study the STT and TTS engines. You implement them in Python to create Melissa's senses. This lays the foundation of how Melissa will interact with you; you use the functionalities implemented in the next chapter throughout the book.

CHAPTER 2

Understanding and Building an Application with STT and TTS

In the previous chapter, you learned to set up the Python project. You learned about three core components that will allow you to interact with Melissa, the virtual smart assistant. In this chapter, you will learn in depth about speech-to-text (STT) and text-to-speech (TTS). You will implement the STT and TTS engines. After this chapter, Melissa will be able to "listen" to your voice input, convert it into text, and use a TTS engine to create a program that repeats the input.

Speech-to-Text Engine

The STT engine is one of the three core components of Melissa. For Melissa to capture the voice input and convert it into text, this engine is crucial. For the purposes of this project, you will be using Vosk. Using Vosk, Melissa will convert the speech into text, in real time.

Vosk

Vosk is a lightweight and efficient STT engine that runs entirely offline. It is well-suited for low-resource devices like the Raspberry Pi, making it an excellent choice for projects requiring voice recognition without an internet connection.

It features the following:

- **Offline processing**: No internet connection required
- **Low resource usage**: Works efficiently on Raspberry Pi
- **Supports multiple languages**: Various pre-trained models are available
- **Real-time transcription**: Can process live audio input

Setting Up Vosk

To install Vosk, run the following Python installation command:

```
pip install vosk
```

Vosk requires a language model to transcribe speech. This book demonstrates how to work with the `vosk-model-small-en-us-0.15` model. However, you can download any model from Vosk and extract it. Install the `vosk-model-small-en-us-0.15` model using the cURL command, then extract the contents using the unzip command, and lastly move to a folder named model.

```
curl -O https://alphacephei.com/vosk/models/vosk-model-small-en-us-0.15.zip
unzip vosk-model-small-en-us-0.15.zip
mv vosk-model-small-en-us-0.15 model
rm vosk-model-small-en-us-0.15.zip
```

CHAPTER 2   UNDERSTANDING AND BUILDING AN APPLICATION WITH STT AND TTS

Now that you have Vosk installed and a model to use with it, let's add some code to the stt.py file.

Implementing STT

Vosk allows you to transcribe the speech in real time. You will add the code that will listen for voice input, transcribe it in real time, and stop if there is no more input. In your stt.py file, add the following code:

```
import sys
import os
from vosk import import Model, KaldiRecognizer, SetLogLevel
import pyaudio
import json
import time

def stt():
    SetLogLevel(-1) # Disable Vosk logging
    # Load the Vosk model
    model = Model("model")
    recognizer = KaldiRecognizer(model, 16000)

    # Start audio stream
    mic = pyaudio.PyAudio()
    stream = mic.open(format=pyaudio.paInt16, channels=1,
    rate=16000, input=True, frames_per_buffer=4096)
    stream.start_stream()

    MAX_SILENCE_TIME = 2.0  # Maximum time to wait for speech
    INITIAL_TIMEOUT = 5.0   # Longer initial timeout to wait
    for speech
    last_speech_time = time.time()
    has_speech = False
    resultText = ""
```

CHAPTER 2 UNDERSTANDING AND BUILDING AN APPLICATION WITH STT AND TTS

```python
    print("Listening...")

while True:
    data = stream.read(4096, exception_on_overflow=False)

    if recognizer.AcceptWaveform(data):
        result = json.loads(recognizer.Result())
        if result['text'].strip():
            has_speech = True
            last_speech_time = time.time()
            resultText += f"{result['text']} "
            print(f"Recognized: {resultText}")
    else:
        # Check partial results for ongoing speech
        partial = json.loads(recognizer.PartialResult())
        if partial.get('partial', '').strip():
            has_speech = True
            last_speech_time = time.time()

    current_time = time.time()
    # Check if we've had speech and now there's a pause
    if has_speech and (current_time - last_speech_time) >
    MAX_SILENCE_TIME:
        print("Speech completed, stopping...")
        break

    # Use longer timeout for initial speech detection
    if not has_speech and (current_time - last_speech_time)
    > INITIAL_TIMEOUT:
        print("No speech detected, stopping...")
        return None

# Clean up
stream.stop_stream()
```

CHAPTER 2   UNDERSTANDING AND BUILDING AN APPLICATION WITH STT AND TTS

```
stream.close()
mic.terminate()

return resultText
```

Let's understand what the code does in detail:

- The code imports all the required packages.

- The code defines an `stt` function.

- Within this function, it loads the model you installed previously, using KaldiRecognier, and sets the encoding to 16000Hz.

- The code turns on the microphone on your machine and starts listening to voice input.

- The code sets certain variables that are used to determine whether there is any voice input.

- The code sets a variable `resultText` that is used to store the transcription.

- In the while loop, the voice input from the user is captured, transcribed, and added to the resultText variable.

- Moreover, in the `while` loop, the code checks if the user is still speaking by checking the partial results from Vosk and also checks for any pauses. Lastly, it gives the user five seconds to start the conversation before the code terminates due to no user input.

- At termination, it stops and closes the stream, and terminates the microphone connection.

- Lastly, it returns the transcription.

CHAPTER 2 UNDERSTANDING AND BUILDING AN APPLICATION WITH STT AND TTS

You can adjust the `frames_per_buffer` value and the value of the first parameter in the `stream.read()` method, if your voice is not correctly recognized. `frames_per_buffer` determines how much audio data is read from the microphone before it is processed. The first parameter in the `stream.read()` method describes the number of audio frames that should be read from the data stream.

You now have a speech-to-text engine that can help you communicate with Melissa. In the next section, you will integrate this with main program and test it.

Melissa's Inception

You have implemented the STT engine, and it's time to test it. After saving the previous code in the `stt.py` file, update the main.py file. This is where you will make use of the core engines.

In your `main.py` file, import the `stt` function and call it.

```
from stt import stt

userInput = stt()
print(userInput)
```

Execute the `main.py` file to the STT engine in action. Make sure your microphone is working before you run the program.

```
python3 src/main.py
```

You should get a similar output in your terminal.

```
Listening...
Recognized: hello and welcome to this book
Speech completed, stopping...
hello and welcome to this book
```

CHAPTER 2   UNDERSTANDING AND BUILDING AN APPLICATION WITH STT AND TTS

With this, the first of three components of the project is complete. You can now speak to your computer and the speech will be converted to a string!

Text-to-Speech Engine

Meliss has got "hearing" capability. However, it cannot still "speak" to you. In this section, you will focus on the second component of the project: converting text to speech. Having a personal assistant speak back creates a pleasant experience for end users, so that's what you're going to implement next.

You can use platforms like ElevenLabs (`elevenlabs.io`) to convert the text into speech. If you use ElevenLabs, you can use a custom voice to make it more personalized. However, using ElevenLabs is not free, and it also means that it would not work offline.

Since the virtual assistant should be able to work offline, you will use the operating system's default TTS engine. In this section, you will try the TTS engine of your operating system—macOS or Linux. The latter will come into play for Raspberry Pi.

macOS

Built into the macOS system is a robust TTS engine that can be customized via its System Settings app as needed. This means you do not need to install any third-party software, and you can start testing the engine in a terminal right away:

```
say "Hi, I am Melissa"
```

Turn on your speakers or try on some headphones, and you'll hear the sentence being spoken out loud!

Additionally, you can set the voice you want to use by adding the -v flag along with the desired voice name. You can find the available voices by going to System Settings ➤ Accessibility ➤ Spoken Content ➤ System voice. Try Samantha with the -v flag. If you have heard of Apple's Siri, Samantha sounds similar to it.

```
say -v Samantha "Hi I am Melissa"
```

If Samantha isn't available on your machine, you will hear a different voice. If you are running an old version of macOS, update your macOS to the latest version to get new voices. At the time of writing this book, I am on macOS Sequoia v15.3.

Linux

Most Linux systems come with a piece of software called *espeak* pre-installed. If it is not, instructions on how to do so can be found at https://espeak.sourceforge.net/. Make sure you add the command to the system path to make it accessible.

Once installed, you can run the following command:

```
espeak "Hi I am Melissa"
```

Your machine will now speak to you! Note that the quality may not be as high as expected from macOS. You can also adjust the speed and make it speak fast or slow, using the -s flag. The default value is 175 (175 words per minute). You can learn more about the -s flag and other flags, in the espeak documentation (https://espeak.sourceforge.net/commands.html).

Like with the macOS say command, you can install and utilize voices using the -v flag. Additional voices can be installed through https://github.com/numediart/MBROLA-voices.

CHAPTER 2 UNDERSTANDING AND BUILDING AN APPLICATION WITH STT AND TTS

Implementing the TTS Engine

The operating system provides a default TTS engine, and this is what you are going to use for Melissa. Based on the operating system the code gets executed on, the TTS engine will be used.

Using Python's sys.platform function, you will determine which operating system is running. The value of sys.platform on Apple systems is Darwin, and on Linux-based systems it is either linux or linux2. Hence, if the operating system is macOS (Darwin), Melissa will use the say command; otherwise, if it is linux (or linux2), it will use espeak. You also use the shelx package that will allow the TTS engine handle nonwords characters like a single quote ('), dash (-), and others.

Your tts.py file should have the following code:

```python
import os
import sys
import shlex

def tts(message):
    """
```

This function takes a message as an argument and converts it to speech depending on the OS.

```
    """

    message = str(message)
    escaped_message = shlex.quote(message)

    if sys.platform == 'darwin':
        tts_engine = 'say'
        return os.system(f'{tts_engine} {escaped_message}')
    elif sys.platform == 'linux2' or sys.platform == 'linux':
        tts_engine = 'espeak'
        return os.system(f'{tts_engine} {escaped_message}')
```

CHAPTER 2   UNDERSTANDING AND BUILDING AN APPLICATION WITH STT AND TTS

You have implemented both the STT and TTS engines for Melissa. You also added the STT engine in the main.py file and tested it. You will now add the TTS engine to the main.py file so that Melissa will greet you and then listen to your input and transcribe it.

Update the main.py file with the following code:

```
from stt import stt
from tts import tts

tts("Hello, I am Melissa. How can I help you?")
userInput = stt()
print(userInput)
```

Execute the program, and you can hear Melissa speak!

EXERCISE: ECHO

Now that you've built both an STT system and a TTS system, how about combining them? Try writing a program that repeats what you say.

Adding a Personal Touch

In the first chapter, you created a file called profile.json. This file contains some information about you, such as your name, city, etc. In this section, you will fetch information from this file and let Melissa use it to greet you. Moreover, Melissa will repeat what you said.

Update the main.py file with the following code:

```
from stt import stt
from tts import tts
import json
```

CHAPTER 2   UNDERSTANDING AND BUILDING AN APPLICATION WITH STT AND TTS

```
# Load the profile.json file
with open("profile.json") as f:
  profile = json.load(f)

name = profile["name"]

tts("Hello " + name + ", I am Melissa. How can I help you?")

userInput = stt()
print(userInput)

tts("You said: " + userInput)
```

Note the changes to the code.

The profile is now loaded and greeted by Melissa using `tts`. The speech is transcribed as before using stt and is now repeated by Melissa once again using tts.

Feel free to commit and push your changes using Git.

Get the Code

If you are running into issues or want to just get the code, you can find it on the GitHub repository: https://github.com/Melissa-AI/melissa-v2. Clone the repository, and check out to the `chapter-2` branch. You can use the code as is. Just make sure to create and activate a virtual environment and install the packages.

```
// clone the repo
git clone https://github.com/Melissa-AI/melissa-v2.git
// navigate into the repo
cd melissa-v2
// switch to chapter 2 branch
git checkout chapter-2
// create the virtual environment
```

```
python3 -m venv venv
// start the virtual environment
source venv/bin/activate
// install packages
pip install -r requirements.txt
```

Summary

In Chapter 1, you created the project structure. In this chapter, you learned more about two of the three core components: STT and TTS. You implemented STT using the Vosk package. This transcribes voice input in real time.

You also learned about the default TTS engines on macOS and Linux. You learned to use the system's built-in functionality in Python by using the operating system's default TTS engine for Melissa. You also integrated these engines into the `main.py` file so that you can interact with Melissa.

In the next chapter, you will learn about the logic engine, how to implement it, and how to integrate the core components together.

CHAPTER 3

Making Melissa Intelligent

In the previous chapter, you added the speech-to-text (STT) and text-to-speech (TTS) functionalities for your virtual assistant Melissa. You can now talk to the assistant, and the assistant repeats the input. The assistant does not really understand your request since you have not implemented the logic engine. In this chapter, you will implement the logic engine and make Melissa intelligent.

Introduction to Large Language Models

Computers only understand binary data (1s and 0s). If you instruct a computer to perform an action, it will first convert it into machine-understandable binary data and look for the instructions on how to perform the action. If the instructions are available, it will perform the action and return the output in a humanly understandable format. This works great if the number of actions you want the computer to perform is limited and does not require the computer to make decisions on its own.

For a virtual assistant, however, you would want it to "think" and make some decisions by itself for drafting an email, posting on social media, checking your calendar for availability and booking meetings for you,

CHAPTER 3   MAKING MELISSA INTELLIGENT

and more. For your assistant to make decisions, you need to use artificial intelligence (AI). The recent developments in this field have made it possible for computers to think and make decisions for humans.

Large language models (LLMs) are AI models that are trained on huge datasets. These models are powerful enough to "think" and generate text, audio, images, and videos. You can use an LLM model to draft an email for you, write poems, or even build a full-stack application. These models are widely being used to accomplish day-to-day tasks.

ChatGPT is one of the most popular examples of the application of an LLM. It is an online chat application that can be used to ask any sort of questions, solve mathematical equations (sometimes it fails to produce correct answers), write code, and much more. However, one should always be careful when using LLMs. They might provide incorrect information. This is widely known as hallucination.

Using an LLM

For your virtual assistant, you will use an LLM that will be the logic engine. The user input in textual format will be passed on to the LLM. The LLM will process the input and return the textual response accordingly. This response will be then passed to the TTS engine, which will speak the response for the user.

CHAPTER 3 MAKING MELISSA INTELLIGENT

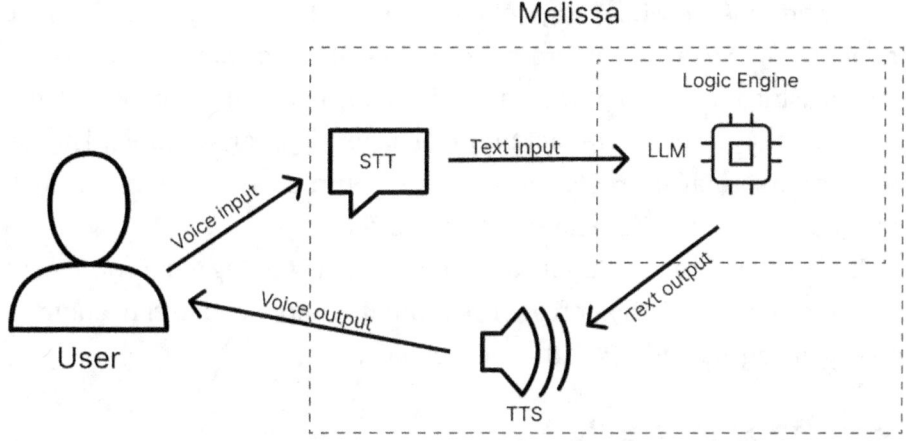

Figure 3-1. *Input-output flow*

There are two different ways to use the LLM for your project. You can either make an API call to the LLM with the user input or run the LLM locally on the machine. Both have their own pros and cons. Let us try to understand them.

Using an API

If you use an LLM hosted on the cloud and make API calls to them, you might get better response and, based on the network connectivity, faster results as well. Since these LLMs are hosted on more powerful computers with more memory, using an API makes sense if you have limited memory (<1 GB).

While this seems like a good solution, there are two key issues. The first is that you cannot run your virtual assistant completely offline. It needs to be connected to the internet at all times. When connecting virtual assistants like Melissa to the internet, it's important to implement appropriate security measures to protect your data and prevent unauthorized access. Basic security practices can help ensure your assistant is used safely for tasks like email and scheduling.

CHAPTER 3   MAKING MELISSA INTELLIGENT

The other issue with using a cloud-based API is privacy. The LLM API providers often use the data from these API calls to re-train the models. If you are using Melissa to process sensitive information like your email or home address, chances are that they can be used to train a model. This can result in your email or home address being returned as a response from the LLM to someone else using the same model.

If you are building an application that is always going to be connected to the internet, needs faster responses, and does not need to care about privacy, using a cloud API is the ideal choice.

Using an LLM locally

For Melissa, you want the data to stay on your computer. While the resulting output might be slow, if the right model is selected, you can get better results. Hence, you will run an LLM locally. This will let Melissa perform tasks without needing to connect to the internet, and your data remains safe.

At the time of authoring this book, there are two popular ways you can run LLMs locally: llamafile and Ollama. Both llamafile and Ollama allow you to install any of the available LLM they support, respectively. But before installing the models, you need to install the software that can run these models. In this book, you will use Ollama.

If you are curious about running llamafile, you can find the instructions on their GitHub repository: `https://github.com/Mozilla-Ocho/llamafile`.

Installing Ollama

To use Ollama, you can download and install it from the Ollama website: `https://ollama.com/download`. Select the operating system you are running and download the installer. Once the installer is downloaded, execute it to install Ollama.

CHAPTER 3 MAKING MELISSA INTELLIGENT

If you are using Linux, you need to execute the following command to install Ollama:

```
curl -fsSL https://ollama.com/install.sh | sh
```

Once Ollama is installed, you can access it in your terminal. Execute the command:

```
ollama
```

You will see a similar output as following:

```
Usage:
  ollama [flags]
  ollama [command]
Available Commands:
    serve       Start ollama
    create      Create a model from a Modelfile
    show        Show information for a model
    run         Run a model
    stop        Stop a running model
    pull        Pull a model from a registry
    push        Push a model to a registry
    list        List models
    ps          List running models
    cp          Copy a model
    rm          Remove a model
    help        Help about any command
Flags:
  -h, --help      help for ollama
  -v, --version   Show version information
Use "ollama [command] --help" for more information about a command.
```

CHAPTER 3   MAKING MELISSA INTELLIGENT

Now that you have Ollama installed, you can use it to run any model mentioned on their website: https://ollama.com/search. For Melissa, you will use the qwen3:1.7b model with three billion parameters. This model also supports function calling. You will learn about this in the next chapter.

To use this model, execute the ollama run command. This will first download the model and then start a local server that will let you interact with the model.

```
ollama run qwen3:1.7b
```

Congratulations! You can now interact with the qwen3 model running locally on your computer.

Use the Local Server

You can interact with the LLM via the terminal. You can also make API calls to a local server. Whenever you run an LLM on Ollama, a local server spins up. Moreover, Ollama provides a Python package that makes it simpler to use in your project. Before you start implementing the Python code, try running a cURL command in your terminal and make a request asking a question. Make sure you still have Ollama running. If it is not running, open the Ollama app you installed, or start the server using the following command:

```
ollama serve
```

This will start the local server on port 11434. You can execute a cURL command and make a post request. By default, the API returns the response as a stream.

```
curl http://localhost:11434/api/chat -d '{
  "model": "qwen3:1.7b",
  "messages": [
```

```
    { "role": "user", "content": "What is a Raspberry Pi?" }
  ]
}'
```

You can set stream to false in the body payload to get the response as a single object. Since the API will return the response as a single object, it might take a few seconds to see the response in the terminal.

```
curl http://localhost:11434/api/chat -d '{
  "model": "qwen3:1.7b",
  "stream": false,
  "messages": [
    { "role": "user", "content": "What is a Raspberry Pi?" }
  ]
}'
```

You can now make calls to the LLM via an API endpoint.

Implement the Logic Engine for Melissa

You have the LLM model running locally on your machine. You can interact with it via the terminal or using the API. In this section, you will learn how to interact with the LLM in your Python code. This implementation will provide the missing piece for Melissa: the logic engine. Melissa will get artificial intelligence that can help you create a true virtual assistant.

To implement this intelligence in your code, you will use the official Python package provided by Ollama (https://github.com/ollama/ollama-python). At the time of writing this book, the OpenAI API is experimental and might change in the future. Hence, you will use the official package from Ollama. This package is a wrapper around Ollama's API.

CHAPTER 3   MAKING MELISSA INTELLIGENT

The first step is to install the package. Use pip to install the required package.

```
pip install ollama
```

Just like you added the code for the TTS engine and the STT engine in their respective files, you will add the code for the logic engine in the logic.py file. Update the logic.py file to import the chat method and the ChatResponse class. You will also update the message passed to the tts method. You will pass the message generated from the LLM to make Melissa's responses more dynamic. Add the following code after the import statements:

```
import re
from ollama import chat, ChatResponse

def logic(message):
    """
    This function takes a user message as an argument and passes
    it to the LLM.
    The response from the LLM is then sent to the tts function.
    """
    response = chat(model='qwen3:1.7b', messages=[
        {
            'role': 'system',
            'content': 'You are a helpful virtual assistant
            named Melissa. Give concise replies.',
        },
        {
            'role': 'user',
            'content': message or 'hi',
        },
    ])
```

CHAPTER 3   MAKING MELISSA INTELLIGENT

```
# Clean the response content
if 'content' in response.message and response.message['content']:
    response.message['content'] = re.sub(r'<think>.*?</think>', '', response.message['content'], flags=re.DOTALL).strip()

return response.message.content ent'], flags=re.DOTALL).strip()
```

You have implemented the logic engine that uses an LLM. You need to update your main.py file to use this logic engine. In your main.py file, update the code to import the logic function. Send a hello message with the name from the `profile.json` file as a parameter for this function. You will then use the output of this function in the `tts` function. This will result in Melissa speaking the welcome message that it generates!

You updated code should be as follows:

```
from stt import stt
from tts import tts
# import the logic function from logic.py
from logic import logic
import json

# Load the profile.json file
with open("profile.json") as f:
    profile = json.load(f)

name = profile["name"]

# Pass the name to the logic function
llm_response = logic(f"Hello, my name is {name}")

# Pass the response from the logic function to the tts function
tts(llm_response)
```

CHAPTER 3 MAKING MELISSA INTELLIGENT

```
userInput = stt()
print(userInput)

tts("You said: " + userInput)
```

If you execute the src/main.py file, you will get a new welcome message from Melissa. However, Melissa still can't reply to your voice input messages. Can you try implementing it yourself? Try solving the exercise and continue reading to find out the solution.

EXERCISE

You have installed an LLM model that can be executed locally. You also created a function that takes an input parameter and passes that to the LLM. The response from the LLM is sent to the tts function, where Melissa speaks the response to you. Try updating the code so that the logic function takes the input from the STT engine.

Interact with Melissa

Right now, if you ask Melissa a question, it will repeat your question rather than providing an answer. Since Melissa now has artificial intelligence, it should be able to use it to answer your questions/requests. For this, you need to update the code. You also want to make sure that Melissa processes any further questions/requests. The program should not end after just one request is processed.

To combine all three core components and keep the program alive, update the code in your main.py file as follows:

```
from stt import stt
from tts import tts
# import the logic function from logic.py
```

```
from logic import logic
import json

# Load the profile.json file
with open("profile.json") as f:
   profile = json.load(f)

name = profile["name"]

# Pass the name to the logic function
llm_response = logic(f"Hello, my name is {name}")

# Pass the response from the logic function to the tts function
tts(llm_response)

while True:
  try:
    # Get user input via voice
    userInput = stt()

    # Check if user input is empty or None
    if not userInput:
      print("No input detected. Ending conversation...")
      break

    # Process user input and respond
    tts(logic(userInput))

except Exception as e:
  print(f"An error occurred: {str(e)}")
  break
```

This change adds a while loop with a try..except block. If there are no inputs from the user, the loop breaks, ending the program. Execute the src/main.py file to see this change in action.

CHAPTER 3   MAKING MELISSA INTELLIGENT

Get the Code

If you quickly want to implement intelligence to Mclissa and do not want to get into the theory right now, you can find the code in this GitHub repository: https://github.com/Melissa-AI/melissa-v2. However, you will still need to install Ollama and download the qwen3:1.7b model. Without these, you will run into errors. Make sure you follow the steps in the "Installing Ollama" section.

You can then clone the repository, start a new virtual environment, and install the packages.

```
// clone the repo
git clone https://github.com/Melissa-AI/melissa-v2.git
// navigate into the repo
cd melissa-v2
// switch to chapter 3 branch
git checkout chapter-3
// create the virtual environment
python3 -m venv venv
// start the virtual environment
source venv/bin/activate
// install packages
pip install -r requirements
```

Summary

While in the previous chapter you added the TTS and STT engines, in this chapter you added the logic engine. You learned about large language models and how to run them locally using Ollama. You added the functionality that lets Melissa use an LLM, making it smarter.

In the next chapter, you will learn about the plugin architecture and how to provide Melissa with more functionality. This architecture will allow you to add functionalities for different tasks that Melissa can do for you, for example, managing your to-do list, checking weather, drafting social media posts, etc. For each task, you will create a separate plugin to make Melissa extensible.

CHAPTER 4

Introducing Plugin Architecture

In the previous chapters, you implemented speech-to-text (STT), text-to-speech (TTS), and the logic engine for Melissa. These allow you to have conversations with Melissa and ask questions. Melissa now has artificial intelligence, but it still cannot perform any actions for you such as maintaining a to-do list or even telling the weather.

In this chapter, you will implement a plugin architecture for Melissa. This pattern will help you write different functionalities for the virtual assistant and easily maintain them. Melissa will use these plugins to perform the relevant action.

What Is a Plugin?

In the context of this book, plugins can be defined as individual functionality that is modular and independent of other plugins. These plugins take input, process it to perform a task, and return output. Since plugins are modular, the logic engine can decide and use the plugins required to perform the action.

CHAPTER 4   INTRODUCING PLUGIN ARCHITECTURE

If you are familiar with function calling in large language models (LLMs), you can also think of plugins as tools. You will pass these tools (plugins) to the LLM. Based on your request, the LLM will decide which tools to use, execute the tools, and generate a response from the value returned by the tools.

Create Your First Plugin

LLMs have a cutoff date. They get trained on data that is available until that cutoff date. You can ask the LLM about its cutoff date. They cannot provide information after the cutoff date and also cannot provide real-time information. LLMs cannot even tell you today's date and the current time. In this section, you will create a plugin that the LLM can use to tell the date and time. Run the program and ask Melissa the current date or time. You will observe that Melissa is not able to tell you the correct date or time.

You will create the plugins in a new folder called `plugins`, inside the `src` directory. Inside the plugins folder, create a `getDateAndTime.py` file. The folder structure of your project should be as follows:

```
src/
plugins/
getDateAndTime.py
...
...
```

The `getDateAndTime.py` file will contain three functions:

`get_current_time`: Returns the current time in the format HH:MM

`get_current_date`: Returns the current date in the format DD/MM/YYYY

CHAPTER 4   INTRODUCING PLUGIN ARCHITECTURE

You can use the following code to add these functions to your file:

```python
from datetime import datetime

def get_current_time():
    """Return the current time in HH:MM format"""
    return f"The current time is {datetime.now().strftime("%H:%M")}"

def get_current_date():
    """Return the current date in DD MMM YYYY format"""
    return f"Today's date is {datetime.now().strftime("%d %b %Y")}"
```

To test the functions, you can call the get_date_time function with the input "What is the time right now?" to get the time, and "What is the date?" to get the date. Add the following code to the same file to test the functions:

```python
print(get_current_date())
print(get_current_time())
```

Execute the file using the following Python command:

```
python3 src/plugins/getDateAndTime.py
```

In your terminal, you will see the time and date.

If you get the output, you will observe that the plugin is working well. However, if you ask Melissa about the date or time, you will not receive the correct response. This is because you have not provided a plugin for Melissa to use.

After you try out the plugin, make sure you remove the print statements that call the functions. You don't need them anymore.

43

CHAPTER 4   INTRODUCING PLUGIN ARCHITECTURE

Function Calling in LLMs

You have a plugin that returns data or time, based on the user's input. Melissa needs access to this plugin to return the date or time, whenever the user asks for it. Once Melissa has access to the plugin, based on the users' input, it can decide if it needs to use the plugin or not. If Melissa thinks that it needs to use the plugin, it will use the plugin, passing the input from the user. The plugin will execute and provide a result to Melissa, and Melissa will use this result to generate a response for the user. This is known as *function calling*. You provide an LLM with some functions that perform certain tasks. LLM decides which function to use (if at all needed) to fulfil the users' requests. See Figure 4-1.

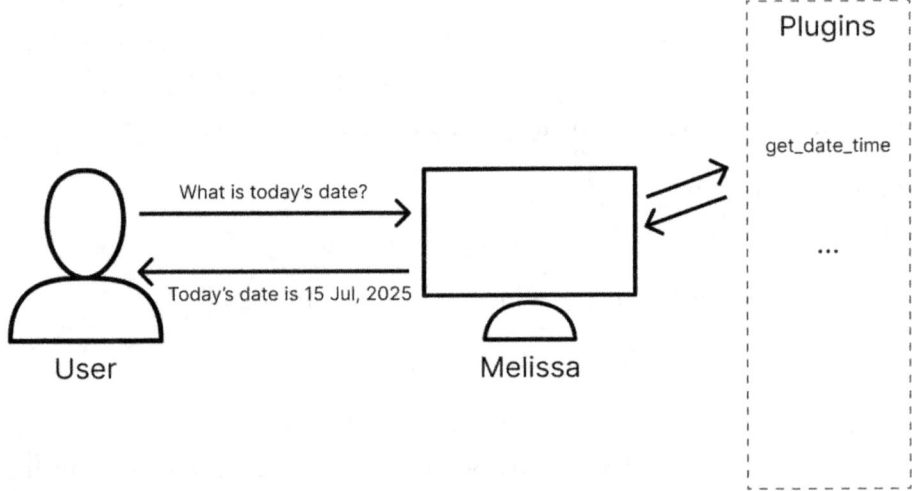

Figure 4-1. Plugins workflow

To provide Melissa with access to the plugin, you need to update the code in the `logic.py` file, as follows:

```
import re
from ollama import chat, ChatResponse
from plugins.getDateAndTime import get_current_date, get_current_time
```

```python
def logic(messages):
    """
    Process messages and handle tool calls
    Returns appropriate response from the LLM.
    """
    available_functions = {
        'get_current_date': get_current_date,
        'get_current_time': get_current_time
    }

    # Define tool schema more explicitly
    tools = [{
        "type": "function",
        "function": {
            "name": "get_current_date",
            "description": "Get today's date",
            "parameters": {
                "type": "object",
                "properties": {},
                "required": []
            }
        }
    }, {
        "type": "function",
        "function": {
            "name": "get_current_time",
            "description": "Get the current time of the day",
            "parameters": {
                "type": "object",
                "properties": {},
                "required": []
            }
```

```python
        }
    }]

    # Get initial response
    response = chat(model='qwen3:1.7b', messages=messages,
    tools=tools)

    # Clean the response content
    if 'content' in response.message and response.
    message['content']:
        response.message['content'] = re.sub(
            r'<think>.*?</think>', '', response.
            message['content'], flags=re.DOTALL).strip()

    # If no tool calls, return the response directly
    if not response.message.tool_calls:
        print('No tool calls detected')
        return response.message

    # Handle tool calls if present
    tool_outputs = []
    for tool in response.message.tool_calls:
        if function_to_call := available_functions.get(tool.
        function.name):
            output = function_to_call(**tool.function.arguments)
            if output:
                tool_outputs.append({
                    'role': 'tool',
                    'content': output,
                    'name': tool.function.name
                })

    if tool_outputs:
        # Add the assistant's tool call message
```

```
    messages.append({
        'role': 'assistant',
        'content': response.message.content or '',
        'tool_calls': response.message.tool_calls
    })
    # Add the tool outputs
    messages.extend(tool_outputs)

    final_response = chat('qwen3:1.7b', messages=messages,
    tools=tools)

    # Clean the final response content
    if 'content' in final_response.message and final_
    response.message['content']:
        final_response.message['content'] = re.sub(
            r'<think>.*?</think>', '', final_response.
            message['content'], flags=re.DOTALL).strip()
    return final_response.message

return response.message
```

You import the function get_current_date and get_current_time from the getDateAndTime.py file. You also update the logic.py file.

Previously, a single message was passed to Melissa. This worked well if it was a one-off conversation. But if a follow-up question was asked, Melissa would not be able to process it. This is because Melissa doesn't have the context of the conversation. The previous code changes that. Instead of passing a message to the logic function, you now pass the messages list.

You also create a dictionary that defines all the available plugins. For now, it only contains get_current_date and get_current_time. When you create more plugins, you should add them to the available_functions dictionary.

CHAPTER 4   INTRODUCING PLUGIN ARCHITECTURE

You also define the function schema of the plugin in the tools list. In the schema, you define the name and description of the plugin. You also specify the parameters that can be passed to the plugin. This provides the LLM with more context on the plugin. You can learn more about it, in the OpenAPI documentation: https://platform.openai.com/docs/guides/function-calling#defining-functions.

You then pass the messages and the tools to the LLM. The LLM processes the input and decides if it needs to use a tool. It uses the tool, if needed, and generates output based on the value returned by the tool. Melissa can now use the plugin to return the current date or time to you.

Before you try it out, you need to update the `main.py` file to accommodate the changes you made to Melissa earlier. Instead of passing a single message, you need to pass a list of messages. This list also contains the system message that helps you define the "personality" of Melissa. The logic function also returns a list of messages; you need to pass only the content of the message to the `tts` function. The following code handles all this:

```
from stt import stt
from tts import tts
from logic import logic
import json

# Load the profile.json file
with open("profile.json") as f:
    profile = json.load(f)

name = profile["name"]

messages = [
    {
        'role': 'system',
        'content':
        '''
```

CHAPTER 4   INTRODUCING PLUGIN ARCHITECTURE

You are an AI assistant designed to be helpful and efficient. You have tools to assist with specific tasks.

TOOL USAGE GUIDELINES:
*   If a user asks a question that your tool is designed to answer, you MUST use the tool.
*   You MUST use your available tools whenever a user's request directly and clearly maps to a tool's capability.

AVAILABLE TOOLS:
   1. **get_current_date:** Get the current date. ***Instruction:** You MUST use this tool for ANY request about the current date (e.g., "What's today's date?"). Do not attempt to answer date questions from your own knowledge.

   2. **get_current_time:** Get the current time. ***Instruction:** You MUST always use this tool for ANY request about the current time (e.g., "What time is it?"). Do not attempt to answer time questions from your own knowledge or conversation history.

For all other questions not covered by your tools, respond naturally.

EXAMPLES:
   User: "What time is it?"
   Assistant: The current time is 12:34

   User: "Tell me today's date"
   Assistant: Today's date is 17 Dec, 2025

   User: "Hello, my name is John"
   Assistant: Hello John, how can I help you today?

CHAPTER 4   INTRODUCING PLUGIN ARCHITECTURE

```
        ...
    },
]

# Initial greeting
messages.append({
    'role': 'user',
    'content': f"Hello, my name is {name}",
})

# Get and speak initial response
llm_response = logic(messages)

tts(llm_response.content)

while True:
    try:
        userInput = stt()
        if not userInput:
            print("No input detected. Ending conversation...")
            break

        # Add previous assistant response and new user input
        messages.append({
            'role': 'assistant',
            'content': llm_response.content
        })
        messages.append({
            'role': 'user',
            'content': userInput,
        })

        # Get and speak new response
        llm_response = logic(messages)
        tts(llm_response.content)
```

```
except Exception as e:
    print(f"An error occurred: {str(e)}")
    break
```

After you update the code, run the program to test the new changes. As in the previous chapter, Melissa will greet you. Try asking Melissa the current date or time. Melissa will now respond with today's time or the current time, respectively!

```
python3 src/main.py
```

Get the Code

If you are running into issues or want to just get the code, you can find it on the GitHub repository: https://github.com/Melissa-AI/melissa-v2. Clone the repository, and check out to the `chapter-4` branch. You can use the code as it is. Just make sure to create and activate a virtual environment and install the packages.

```
// clone the repo
git clone https://github.com/Melissa-AI/melissa-v2.git
// navigate into the repo
cd melissa-v2
// switch to chapter 4 branch
git checkout chapter-4
// create the virtual environment
python3 -m venv venv
// start the virtual environment
source venv/bin/activate
// install packages
pip install -r requirements.txt
```

CHAPTER 4   INTRODUCING PLUGIN ARCHITECTURE

Summary

In this chapter, you learned about the plugin architecture for your virtual assistant, Melissa. You created a plugin that returns the current date and time. You also learned about function calling in LLMs. You then updated the code to provide Melissa with access to the plugin you created. Lastly, you updated the logic function to maintain the conversation context. The date and time plugin uses the system APIs to return the date or time. In the next chapter, you will create another plugin that will use an API on the internet that returns to the top tech stories. You will build a plugin for fetching top stories from Hacker News.

CHAPTER 5

Get Top Tech Stories from Hacker News

In the morning, when having your coffee or breakfast, you want to learn about the new and exciting things in the technology world. If you ask Melissa about this, Melissa won't be able to provide you with the recent top stories from the tech world.

In the previous chapter, you learned about plugin architecture and function calling. In this chapter, you'll learn how to create a plugin that fetches top stories from Hacker News (https://news.ycombinator.com/). Hacker News is a social news website that features content from the tech industry, including articles, discussions, and job postings. Using the plugin Melissa will be able to tell you about the latest trending stories on Hacker News.

This can be a part of your morning routine. While sipping a cup of coffee or having your breakfast, you can ask Melissa for the top stories from Hacker News.

What Is an API?

An application programming interface (API) is a connection between different computers and computer programs. APIs allow you to interact with another computer or computer program and use the provided resources. Using an API, you can build applications that can fetch data from the internet (e.g., your social media posts) or even create new data.

Some APIs are available without any authentication. You don't need any API credentials to access them. While some APIs need credentials to access them securely. The Hacker News API doesn't require any authentication.

Hacker News provides a free, public API that doesn't require authentication. The API provides access to all the content on Hacker News (https://github.com/HackerNews/API). The base URL for the API is https://hacker-news.firebaseio.com/v0.

The API provides several endpoints, like these:

> /topstories.json: Returns the recent top stories

> /item/{id}.json: Returns details about a specific item (story, comment, job, etc.)

Creating the Hacker News Plugin

To create this plugin, create a new file called getHackerNews.py in the plugins directory. You will need to import the required packages and define methods as in the following code:

```
import json
from urllib import request, error
from typing import List, Dict, Optional

BASE_URL = "https://hacker-news.firebaseio.com/v0"

def fetch_item(item_id: int) -> Optional[Dict]:
    """Fetch a single item from Hacker News API"""
    try:
        url = f"{BASE_URL}/item/{item_id}.json"
        with request.urlopen(url) as response:
            return json.loads(response.read())
```

CHAPTER 5   GET TOP TECH STORIES FROM HACKER NEWS

```python
    except (error.URLError, json.JSONDecodeError):
        return None

def fetch_top_stories(limit: int = 5) -> List[Dict]:
    """Fetch top stories from Hacker News"""
    try:
        # Get top stories IDs
        url = f"{BASE_URL}/topstories.json"
        with request.urlopen(url) as response:
            story_ids = json.loads(response.read())

        # Fetch details for each story
        stories = []
        for story_id in story_ids[:limit]:
            story = fetch_item(story_id)
            if story:
                stories.append(story)

        return stories
    except (error.URLError, json.JSONDecodeError):
        return []

def get_hackernews_info(query: str) -> str:
    """Process user query and return Hacker News information"""
    # Default to 5 stories if no number specified
    limit = 5

    # Check if user specified a number of stories
    words = query.lower().split()
    try:
        if 'top' in words and len(words) > words.index('top') + 1:
            limit = int(words[words.index('top') + 1])
            limit = min(limit, 20)  # Cap at 20 stories
```

55

```
        except ValueError:
            pass

    stories = fetch_top_stories(limit)
    if not stories:
        return "Sorry, I couldn't fetch stories from
        Hacker News"

    response = f"Top {len(stories)} stories from Hacker
    News:\n\n"
    for i, story in enumerate(stories, 1):
        title = story.get('title', 'No title')
        author = story.get('by', 'unknown')

        response += f"{i}. {title}\n"
        response += f"   Author: {author}\n"

    return response
        print(get_hackernews_info("top 5 stories"))
```

Let's break down the code:

You import the necessary modules:

> json for parsing JSON responses
>
> urllib for making HTTP requests
>
> typing for type hints

You define the base URL for the Hacker News API.

The `fetch_item` function takes an item ID and fetches its details from the API. It handles potential errors and returns None if the request fails.

The `fetch_top_stories` function:

> Fetches the list of IDs of the recent top stories
>
> For each ID, fetches the story details using fetch_item
>
> Returns a list of story dictionaries

CHAPTER 5   GET TOP TECH STORIES FROM HACKER NEWS

The get_hackernews_info function:

Takes a user query as input

Extracts the number of stories requested (defaults to 5)

Caps the number of stories at 20 to prevent too many API calls

Returns a formatted string with the stories

The print statement calls the get_hackernews_info function with the parameter "top 5 stories."

To test the code, execute getHackerNews.py with the following command:

```
python3 src/plugins/getHackerNews.py
```

You should get similar output, as shown here:
Top 5 stories from Hacker News:

1. OpenAI adds MCP support to Agents SDK
   Author: gronky_
2. Debian bookworm live images now reproducible
   Author: bertman
3. A love letter to the CSV format
   Author: Yomguithereal
4. Malware found on NPM infecting local package with reverse shell
   Author: gnabgib
5. Good-bye core types; Hello Go as we know and love it
   Author: ingve

Make sure you remove the last print statement from the getHackerNews.py file.

57

CHAPTER 5   GET TOP TECH STORIES FROM HACKER NEWS

Integrating the Plugin with Melissa

Now that you have created the Hacker News plugin, you need to integrate it with Melissa. As you did in the previous chapter, you need to update two files.

First, update `logic.py` to include the new plugin. Import the plugin and update `available_functions` and tools, like so:

```
...
from plugins.getHackerNews import get_hackernews_info

def logic(messages):
    """
    Process messages and handle tool calls
    Returns appropriate response from the LLM.
    """
    available_functions = {
        'get_current_date': get_current_date,
        'get_current_time': get_current_time,
        'get_hackernews_info': get_hackernews_info
    }

    # Define tool schema more explicitly
    tools = [{
        "type": "function",
        "function": {
            "name": "get_current_date",
            "description": "Get today's date",
            "parameters": {
                "type": "object",
                "properties": {},
                "required": []
            }
        }
    }
```

```
    }, {
        "type": "function",
        "function": {
                "name": "get_current_time",
            "description": "Get the current time of the day",
                "parameters": {
                    "type": "object",
                    "properties": {},
                    "required": []
                }
        }
    },
        {
        "type": "function",
        "function": {
            "name": "get_hackernews_info",
            "description": "Get top stories from HackerNews",
            "parameters": {
                "type": "object",
                "properties": {
                    "query": {
                        "type": "string",
                        "description": "The user's query about
                        HackerNews stories"
                    }
                },
                "required": ["query"]
            }
        }
    }]
...
```

CHAPTER 5   GET TOP TECH STORIES FROM HACKER NEWS

Next, update the system prompt in main.py, as follows:

```
messages = [
    {
        'role': 'system',
        'content':
        '''
        You are an AI assistant designed to be helpful and
        efficient. You have tools to assist with specific tasks.

        **TOOL USAGE GUIDELINES:**
        *   If a user asks a question that your tool is designed
            to answer, you MUST use the tool.
        *   You MUST use your available tools whenever a user's
            request directly and clearly maps to a tool's
            capability.

        AVAILABLE TOOLS:
            1. **get_current_date:** Get the current date.
            ***Instruction:** You MUST use this tool for ANY
            request about the current date (e.g., "What's
            today's date?"). Do not attempt to answer date
            questions from your own knowledge.

            2. **get_current_time:** Get the current time.
            ***Instruction:** You MUST always use this tool for
            ANY request about the current time (e.g., "What time
            is it?"). Do not attempt to answer time questions
            from your own knowledge or conversation history.

            3. **get_hackernews_info:** Get top stories from
            HackerNews.
            ***Instruction:** You MUST use this tool for ANY
            request about the top stories on HackerNews (e.g.,
```

"What are the top stories on HackerNews?"). Do not attempt to answer HackerNews questions from your own knowledge.

For all other questions not covered by your tools, respond naturally.

```
EXAMPLES:
    User: "What time is it?"
    Assistant: The current time is 12:34

    User: "Tell me today's date"
    Assistant: Today's date is 17 Dec, 2025

    User: "What are the top stories on HackerNews?"
    Assistant: Here are the top stories from HackerNews:
        1. [Story Title]
        Author: [Author Name]
        2. [Story Title]
        Author: [Author Name]

    User: "Hello, my name is John"
    Assistant: Hello John, how can I help you today?
    '''
},
]
```

Execute the `main.py` file and try asking Melissa for the top stories from Hacker News.

`python3 src/main.py`

Melissa will use the plugin to fetch the top stories from the Hacker News API and share them with you.

CHAPTER 5   GET TOP TECH STORIES FROM HACKER NEWS

Get the Code

You can find the final code for this chapter, along with the previous chapters in the GitHub repository: https://github.com/Melissa-AI/melissa-v2. Clone the repository, and check out to the chapter-5 branch. Make sure to create and activate a virtual environment and install the packages.

Summary

In this chapter, you learned about APIs and how to interact with an API. You created a plugin that interacts with the Hacker News API. You created functions to fetch top stories and their details and integrated the plugin with Melissa.

In the morning, when sipping coffee, you can ask Melissa for top stories from Hacker News, and Melissa will use this plugin to provide you with a response. But if you ask Melissa if you need an umbrella when heading out, Melissa won't be able to help! In the next chapter, you will build a plugin to fetch the weather forecast.

CHAPTER 6

Mellisa, Tell Me the Weather!

In the previous chapter, you learned about APIs and created a plugin that fetches information from an API. In this chapter, you will create a plugin that will use a weather API to fetch the current weather in the city you specify. You can use this plugin to decide if you need a jacket for the cold winds or an umbrella for showers.

To get the current weather information, you will use the OpenWeather (`https://openweathermap.org/`) API. Unlike the Hacker News API, this API needs to be authenticated. Without authentication, you can't make successful API calls to the OpenWeather APIs. OpenWeather uses an API key to allow you to make authenticated requests.

Get OpenWeather API Key

There are different ways an API can implement authentication; an API key is one of the most popular approaches. An API key is a uniquely generated string that you use to call the API. Each user has their own unique API key, and this helps the API provider (OpenWeather in this case) identify the user making the API call. So whenever you use an API key, make sure you don't share it with anyone. In the later sections, you will learn the best practices to securely use API keys.

CHAPTER 6   MELLISA, TELL ME THE WEATHER!

OpenWeather provides a bunch of APIs like the Air Pollution API (https://openweathermap.org/api/air-pollution), Geocoding API (https://openweathermap.org/api/geocoding-api), Current Weather API (https://openweathermap.org/current), and more. Since you want Melissa to tell you the current weather, you will use the Current Weather API. At the time of writing this book, this API is available in the free tier, and you don't need to pay or share your credit card information.

To get started with the API, you need an OpenWeather account. If you don't have an account, you can sign up for a new account on their website (https://home.openweathermap.org/users/sign_up). On the sign-up page, you might be asked to enter a username, email address, and password. Once the account is created successfully, you will receive a verification email. Make sure you follow the instructions in the email to verify your account.

After verification, you will receive another email. This email will contain the API key that you will need in the next steps. So, make sure to save it somewhere safe. Do not share this API key with anyone. Someone with access to your API key can make a request to the API on your behalf.

Let's test it! Using cURL from your terminal, you will make a request to the API. Make sure to edit the command to add the name of your city and your API key to get the current weather for your city.

```
curl "http://api.openweathermap.org/data/2.5/weather?q=CITY&appid=API_KEY&units=metric"
```

You should get similar output, as shown here:

```
{"coord":{"lon":13.4105,"lat":52.5244},"weather":[{"id":800,"main":"Clear","description":"clear sky","icon":"01d"}],"base":"stations","main":{"temp":14.55,"feels_like":13.24,"temp_min":12.83,"temp_max":15.6,"pressure":1022,"humidity":45,"sea_level":1022,"grnd_level":1016},"visibility":10000,"wind":{"speed":0.89,"deg":134,"gust":3.13},"clouds":{"all":7},"dt":
```

CHAPTER 6   MELLISA, TELL ME THE WEATHER!

1742548685,"sys":{"type":2,"id":2011538,"country":"DE","sunrise":1742533591,"sunset":1742577629},"timezone":3600,"id":2950159,"name":"Berlin","cod":200}%

Create the Weather Plugin

You now have access to the API that returns the current weather of your city. You need to create a plugin and provide Melissa the access to this plugin. Once Melissa has access to the weather plugin, you will be able to ask Melissa about the current weather!

Create a `getWeather.py` file inside the `plugins` directory. Add the following code to get the current weather for a city.

```python
import json
from urllib import request, parse, error
from typing import Dict, Optional

# You'll need to sign up at OpenWeatherMap and get an API key
API_KEY = "YOUR_API_KEY"
BASE_URL = http://api.openweathermap.org/data/2.5/weather

def get_weather(city: str) -> str:
    """Fetch weather data for given city from
       OpenWeatherMap API"""
    try:
        params = {
            'q': city,
            'appid': API_KEY,
            'units': 'metric'  # For Celsius
        }
        url = f"{BASE_URL}?{parse.urlencode(params)}"
        with request.urlopen(url) as response:
```

```python
            weather_data = json.loads(response.read())
            if not weather_data:
                return f"Sorry, I couldn't fetch weather
                information for {city}"

            temp = weather_data['main']['temp']
            description = weather_data['weather'][0]
            ['description']
            humidity = weather_data['main']['humidity']

            return f"Current weather in {city.title()}:\n" \
                f"Temperature: {temp}°C\n" \
                f"Conditions: {description.capitalize()}\n" \
                f"Humidity: {humidity}%"\
    except (error.URLError, json.JSONDecodeError):
        return None
print(get_weather("Berlin"))  # Test the function with a sample query
```

The previous code imports all the required packages and defines the API key and the API endpoint. For now, replace YOUR_API_KEY with your OpenWeatherMap API key. In the next step, you will learn how to securely store the API key. For testing, you can paste the API key in the code.

You define a get_weather function that takes the city as the input parameter and makes a call to the OpenWeatherMap API. It uses your API key to make an authenticated request to the API and returns the weather data.

To test this plugin, execute the getWeather.py file.

```
$ python3 src/plugins/getWeather.py
```

You should get similar output, as shown here:

```
Current weather in Berlin:
Temperature: 10.31°C
Conditions: Clear sky
Humidity: 47%
```

Melissa, What Is the Temperature?

You have a weather plugin that returns the weather information. However, if you ask Melissa about the current weather, the virtual assistant will not be able to give you a correct response. This is because you have not provided this plugin to Melissa yet.

In the previous chapters, you set up the foundation of the function calling LLM and added the Hacker News plugin. To add new tools, you need to make minor changes.

First, update the `logic.py` file. Import the `get_weather` plugin and add it to the `available_functions` and tools list.

```
...
from plugins.getWeather import get_weather

def logic(messages):
    """
    Process messages and handle tool calls
    Returns appropriate response from the LLM.
    """
    available_functions = {
        'get_date_time': get_date_time,
        'get_hackernews_info': get_hackernews_info,
        'get_weather': get_weather,
    }
```

```python
# Define tool schema more explicitly
tools = [{
    "type": "function",
    "function": {
            "name": "get_current_date",
        "description": "Get today's date",
        "parameters": {
                    "type": "object",
                    "properties": {},
                    "required": []
        }
    }
}, {
    "type": "function",
    "function": {
            "name": "get_current_time",
        "description": "Always get the current time of the day",
            "parameters": {
                "type": "object",
                "properties": {},
                "required": []
            }
    }
},
    {
    "type": "function",
    "function": {
        "name": "get_hackernews_info",
        "description": "Get top stories from HackerNews",
        "parameters": {
```

```
            "type": "object",
            "properties": {
                "query": {
                    "type": "string",
                    "description": "The user's query about
                    HackerNews stories"
                }
            },
            "required": ["query"]
        }
    }
},
{
    "type": "function",
    "function": {
        "name": "get_weather",
        "description": "Always get weather information for a
        specific city",
        "parameters": {
            "type": "object",
            "properties": {
                "city": {
                    "type": "string",
                    "description": "The name of the city to
                    get weather information for"
                }
            },
            "required": ["city"]
        }
    }
}]
```

CHAPTER 6   MELLISA, TELL ME THE WEATHER!

```
    # Get initial response
    ...
```

Next, update the system prompt in the main.py file. You should add the plugin under the AVAILABLE TOOLS section and also add an example, as shown here:

```
messages = [
    {
        'role': 'system',
        'content':
        '''
```

You are an AI assistant designed to be helpful and efficient. You have tools to assist with specific tasks.

TOOL USAGE GUIDELINES:
*   If a user asks a question that your tool is designed to answer, you MUST use the tool.
*   You MUST use your available tools whenever a user's request directly and clearly maps to a tool's capability.

AVAILABLE TOOLS:
   1. **get_current_date:** Get the current date.
   ***Instruction:** You MUST use this tool for ANY request about the current date (e.g., "What's today's date?"). Do not attempt to answer date questions from your own knowledge.

   2. **get_current_time:** Get the current time.
   ***Instruction:** You MUST always use this tool for ANY request about the current time (e.g., "What time is it?"). Do not attempt to answer time questions from your own knowledge or conversation history.

CHAPTER 6   MELLISA, TELL ME THE WEATHER!

3. **get_hackernews_info:** Get top stories from HackerNews.
***Instruction:** You MUST use this tool for ANY request about the top stories on HackerNews (e.g., "What are the top stories on HackerNews?"). Do not attempt to answer HackerNews questions from your own knowledge.

4. **get_weather:** Get weather information for a specific city.
***Instruction:** You MUST use this tool for ANY request about the weather in a specific city (e.g., "What's the weather in London?"). Do not attempt to answer weather questions from your own knowledge.

For all other questions not covered by your tools, respond naturally.

EXAMPLES:
   User: "What time is it?"
   Assistant: The current time is 12:34

   User: "Tell me today's date"
   Assistant: Today's date is 17 Dec, 2025

   User: "What are the top stories on HackerNews?"
   Assistant: Here are the top stories from HackerNews:
      1. [Story Title]
      Author: [Author Name]
      2. [Story Title]
      Author: [Author Name]

   User: "What's the weather in London?"
   Assistant: Here's the current weather in London:

71

```
            Conditions: Partly cloudy
            Temperature: 12 C
            Humidity: 65%

        User: "Hello, my name is John"
        Assistant: Hello John, how can I help you today?
    '''
    },
]
```

Lastly, you can remove the print statement in the `src/plugins/getWeather.py` file that was added to test the plugin. You do not need to call the get_weather function inside `getWeather.py` anymore.

Start the Ollama server and execute the `main.py` file. Ask Melissa about the temperature in your city, and you will hear a response from Melissa with accurate information.

```
$ ollama serve
$ python3 src/main.py
```

Securely Storing the API Key

Your virtual assistant can make API calls to the Current Weather API and fetch information. The API uses an API key for a secure call. You have hard-coded the API key in the `getWeather.py` file, which is not secure. In this section, you will learn about securely storing and using API keys and other credentials.

The best practice to store the API keys and other secret credentials is to use an environment variable file. You will create a `.env` file. You will add this file to `.gitignore` to make sure you don't accidentally commit this file and leak your credentials. To do this, enter the following commands in your terminal:

CHAPTER 6   MELLISA, TELL ME THE WEATHER!

```
$ touch .env
$ echo ".env" >> .gitignore
```

The updated `.gitignore` file should contain the following content:

```
venv/
__pycache__/
*.pyc
.env
```

The `.env` file should have the following content. Make sure to replace YOUR_API_KEY with your OpenWeatherMap API key.

```
WEATHER_API_KEY = YOUR_API_KEY
```

You will load the API key from the `.env` file. You will need the `dotenv` package to load the data from this file. Execute the following command to install the package:

```
$ pip install python-dotenv
```

Using this package, you can now securely load the API key in the `getWeather.py` file. Import the package in the `getWeather.py` file and update the code to load the API key from the `.env` file. You should also remove the hard-coded API key from this file since it will now get loaded from the `.env` file.

```
import json
from urllib import request, parse, error
from typing import Dict, Optional
import os
from dotenv import load_dotenv

# Load environment variables from .env file
load_dotenv()
```

CHAPTER 6   MELLISA, TELL ME THE WEATHER!

```python
# You'll need to sign up at OpenWeatherMap and get an API key
# Get API key from environment variable
API_KEY = os.environ.get('WEATHER_API_KEY')

if not API_KEY:
    raise ValueError("API key for OpenWeatherMap not found")

BASE_URL = "http://api.openweathermap.org/data/2.5/weather"

def get_weather(city: str) -> str:
    """Fetch weather data for given city from
        OpenWeatherMap API"""
    try:
        params = {
            'q': city,
            'appid': API_KEY,
            'units': 'metric'  # For Celsius
        }
        url = f"{BASE_URL}?{parse.urlencode(params)}"
        with request.urlopen(url) as response:
            weather_data = json.loads(response.read())
            if not weather_data:
                return f"Sorry, I couldn't fetch weather
                information for {city}"

            temp = weather_data['main']['temp']
            description = weather_data['weather'][0]
            ['description']
            humidity = weather_data['main']['humidity']

            return f"Current weather in {city.title()}:\n" \
                f"Temperature: {temp}°C\n" \
                f"Conditions: {description.capitalize()}\n" \
                f"Humidity: {humidity}%"\
```

CHAPTER 6   MELLISA, TELL ME THE WEATHER!

```
except (error.URLError, json.JSONDecodeError):
    return None
```

Execute the program and try asking Melissa about the temperature in your city. You will notice that Melissa is still able to tell you the correct temperature. The only thing that has changed is how the API key is accessed. It is now more secure than it was before.

EXERCISE

You can now ask Melissa about the current weather, and Melissa will use the plugins you created to provide the information. While the information returned by the get_weather plugin is good, it can be better. The API provides more information like the minimum temperature, maximum temperature, and feels like temperature (defined as the human perception of weather). Update the plugin to return this information as well. You can refer to the Current Weather API documentation (https://openweathermap.org/current) to learn more.

Solution

To return the minimum, maximum, and feels like temperature, you need to update the getWeather.py file. Update get_weather to get these values from the API response.

```
def get_weather(city: str) -> str:
    """Fetch weather data for given city from
       OpenWeatherMap API"""

    try:
        params = {
            'q': city,
            'appid': API_KEY,
```

```python
            'units': 'metric'  # For Celsius
        }
        url = f"{BASE_URL}?{parse.urlencode(params)}"
        with request.urlopen(url) as response:
            weather_data = json.loads(response.read())
            if not weather_data:
                return f"Sorry, I couldn't fetch weather  
                information for {city}"

            temp = weather_data['main']['temp']
            description = weather_data['weather'][0]
            ['description']
            humidity = weather_data['main']['humidity']
            min_temp = weather_data['main']['temp_min']
            max_temp = weather_data['main']['temp_max']
            feels_like = weather_data['main']['feels_like']

            return f"Current weather in {city.title()}:\n" \
                f"Temperature: {temp}°C\n" \
                f"Conditions: {description.capitalize()}\n" \
                f"Humidity: {humidity}%"\
                f"Min Temp: {min_temp}°C\n" \
                f"Max Temp: {max_temp}°C\n" \
                f"Feels Like: {feels_like}°C"
    except (error.URLError, json.JSONDecodeError):
        return None
```

You should also update the system prompt in the `main.py` file.

```python
messages = [
    {
        'role': 'system',
        'content':
        '''
```

CHAPTER 6 MELLISA, TELL ME THE WEATHER!

You are an AI assistant designed to be helpful and efficient. You have tools to assist with specific tasks.

TOOL USAGE GUIDELINES:
*   If a user asks a question that your tool is designed to answer, you MUST use the tool.
*   You MUST use your available tools whenever a user's request directly and clearly maps to a tool's capability.

AVAILABLE TOOLS:
   1. **get_current_date:** Get the current date.
   ***Instruction:** You MUST use this tool for ANY request about the current date (e.g., "What's today's date?"). Do not attempt to answer date questions from your own knowledge.

   2. **get_current_time:** Get the current time.
   ***Instruction:** You MUST always use this tool for ANY request about the current time (e.g., "What time is it?"). Do not attempt to answer time questions from your own knowledge or conversation history.

   3. **get_hackernews_info:** Get top stories from HackerNews.
   ***Instruction:** You MUST use this tool for ANY request about the top stories on HackerNews (e.g., "What are the top stories on HackerNews?"). Do not attempt to answer HackerNews questions from your own knowledge.

CHAPTER 6   MELLISA, TELL ME THE WEATHER!

    4. **get_weather:** Get weather information for a specific city.
    ***Instruction:** You MUST use this tool for ANY request about the weather in a specific city (e.g., "What's the weather in London?"). Do not attempt to answer weather questions from your own knowledge.

For all other questions not covered by your tools, respond naturally.

EXAMPLES:
    User: "What time is it?"
    Assistant: The current time is 12:34

    User: "Tell me today's date"
    Assistant: Today's date is 17 Dec, 2025

    User: "What are the top stories on HackerNews?"
    Assistant: Here are the top stories from HackerNews:
        1. [Story Title]
        Author: [Author Name]
        2. [Story Title]
        Author: [Author Name]

    User: "What's the weather in London?"

    Assistant: Here's the current weather in London:
        Conditions: Partly cloudy
        Temperature: 12 C
        Humidity: 65%
        Min Temp: 4 C
        Max Temp: 14 C

CHAPTER 6   MELLISA, TELL ME THE WEATHER!

```
        Feels Like: 11 C

        User: "Hello, my name is John"
        Assistant: Hello John, how can I help you today?
    ...
    },
]
```

Now when you ask Melissa about the weather, you will get a better result, which might help you decide if you need that jacket!

Get the Code

If you are running into issues or want to just get the code, you can find it on the GitHub repository: https://github.com/Melissa-AI/melissa-v2. Clone the repository, and check out to the chapter-6 branch. You can use the code as is. Just make sure to create and activate a virtual environment and install the packages. You will also need the API key from OpenWeatherMap.

```
// clone the repo
$ git remote add origin https://github.com/Melissa-AI/melissa-v2.git
// navigate into the repo
$ cd melissa-v2
// switch to chapter 6 branch
$ git checkout chapter-6
// rename .env.copy
$ mv .env.copy .env
// update the .env file. Replace YOUR_API_KEY with your API key
// create the virtual environment
$ python3 -m venv venv
```

79

CHAPTER 6   MELLISA, TELL ME THE WEATHER!

```
// start the virtual environment
$ source venv/bin/activate
// install packages
$ pip install -r requirements.txt
```

Summary

In this chapter, you learn about authenticated APIs and how to interact with them. You created a plugin that fetches weather information from the internet by calling the Current Weather API from OpenWeatherMap. You then updated the code to provide Melissa with access to the plugin you created. Melissa can tell you the correct date, time, and weather!

Now before you head out, you can ask Melissa for weather information that can help you decide if you need an umbrella or not.

You have three plugins now, which get the current date and time, get the top stories from Hacker News, and get weather information. Let's create another plugin that will help you create and save notes. You would be able to use this plugin to store the URL of the Hacker News article you are interested in reading later or add a note for the amazing idea you just had!

CHAPTER 7

Saving Notes with Melissa

In the previous chapters, you learned about APIs and created plugins that fetch information from external sources. In this chapter, you will create a plugin that allows Melissa to save, retrieve, and manage notes locally. This plugin will help you keep track of important information, ideas, or reminders without leaving your conversation with Melissa.

You will be able to use this plugin to save the URL of the Hacker News story you enjoyed, or you can save it to read later. The plugin will save these notes locally, without any information leaving the machine.

Understanding Local Data Storage

There are two ways to store your notes: on a database on the internet or locally on your machine. In the previous chapters, you learned the importance of providing the assistant with limited access to the internet. Besides compromising the data, the other disadvantage of storing the data on the internet is speed. To manage your notes, Melissa will have to access the internet, which introduces network latencies and therefore reduces speed.

CHAPTER 7   SAVING NOTES WITH MELISSA

A much better approach is to save the data locally. When working with local data storage, you have several options:

- **File-based storage**: Storing data in individual files (like JSON, CSV, or text files)
- **Database storage**: Using a database system to store and manage data

For your notes plugin, you will use SQLite, a lightweight database that's perfect for local applications. SQLite stores all data in a single file and provides powerful querying capabilities, making it ideal for managing notes.

Why SQLite?

You can save the data in a JSON file. However, besides being lightweight, SQLite has other advantages:

Data integrity: SQLite provides ACID properties (atomicity, consistency, isolation, durability), which help prevent data corruption.

Efficient querying: With SQLite, you can perform complex queries to search, filter, and sort notes without loading all files into memory.

Concurrent access: SQLite handles multiple readers and writers more efficiently than file-based storage.

Reduced file system overhead: Instead of managing many small JSON files, you have a single database file.

Built-in data validation: SQLite's schema enforcement helps maintain data consistency.

CHAPTER 7   SAVING NOTES WITH MELISSA

Creating the Notes Plugin

Now that you have a better understanding of the advantages of using SQLite, let's create a new plugin that will allow Melissa to save, retrieve, list, update, delete, and search notes.

First, create a new file called notesManager.py in the src/plugins directory. This file will contain all the functionality needed to manage notes.

```
import os
import sqlite3
from datetime import datetime

# Define the directory for storing the database
DB_DIR = os.path.join(os.path.dirname(os.path.dirname(os.path.dirname(os.path.abspath(__file__)))), "data")
os.makedirs(DB_DIR, exist_ok=True)
DB_PATH = os.path.join(DB_DIR, "notes.db")

# Initialize the database
def init_db():
    """Initialize the SQLite database with the notes table"""
    conn = sqlite3.connect(DB_PATH)
    cursor = conn.cursor()

    # Create notes table if it doesn't exist
    cursor.execute('''
    CREATE TABLE IF NOT EXISTS notes (
        id INTEGER PRIMARY KEY AUTOINCREMENT,
        title TEXT NOT NULL,
        content TEXT NOT NULL,
        created_at TEXT NOT NULL,
        updated_at TEXT NOT NULL
    )
    ''')
```

CHAPTER 7   SAVING NOTES WITH MELISSA

```
    conn.commit()
    conn.close()

# Initialize the database when the module is imported
init_db()
```

This code sets up the SQLite database and creates a table to store your notes. Each note will have an ID, title, content, creation timestamp, and last update timestamp.

Next, add the functions to save, retrieve, list, update, delete, and search notes.

Save Note

Here's the code for saving a note:

```
def save_note(title, content):
    """Save a note with the given title and content"""
    conn = sqlite3.connect(DB_PATH)
    cursor = conn.cursor()

    # Check if a note with this title already exists
    cursor.execute("SELECT id FROM notes WHERE title = ?",
    (title,))
    existing_note = cursor.fetchone()

    current_time = datetime.now().isoformat()

    if existing_note:
        # Update existing note
        cursor.execute(
            "UPDATE notes SET content = ?, updated_at = ? WHERE
            title = ?",
            (content, current_time, title)
        )
```

```
            conn.commit()
            conn.close()
            return f"Note '{title}' updated successfully."
        else:
            # Insert new note
            cursor.execute(
                "INSERT INTO notes (title, content, created_at,
                updated_at) VALUES (?, ?, ?, ?)",
                (title, content, current_time, current_time)
            )
            conn.commit()
            conn.close()
            return f"Note '{title}' saved successfully."
```

This function does the following:

1. Get the existing note with the title

2. If the note exists, updates the existing note with the new content

3. If the note doesn't exist, adds the note to the table

Get Note

Here's the code for fetching a note:

```
def get_note(title):
    """Retrieve a note by its title"""
    conn = sqlite3.connect(DB_PATH)
    cursor = conn.cursor()

    cursor.execute("SELECT title, content, created_at FROM
    notes WHERE title = ?", (title,))
    note = cursor.fetchone()
```

```
    conn.close()

    if not note:
        return f"No note found with title '{title}'."

    title, content, created_at = note
    return f"Note: {title}\nCreated: {created_at}\n\n{content}"
```

The function fetches the note with the title. If the note is not present, it returns a "not found" message; otherwise, it returns the information of the note.

List Notes

Here's the code for listing notes:

```
def list_notes():
    """List all available notes"""
    conn = sqlite3.connect(DB_PATH)
    cursor = conn.cursor()

    cursor.execute("SELECT title, created_at FROM notes ORDER BY created_at DESC")
    notes = cursor.fetchall()

    conn.close()

    if not notes:
        return "No notes found."

    notes_list = []
    for title, created_at in notes:
        # Format the date to be more readable
        date_obj = datetime.fromisoformat(created_at)
        formatted_date = date_obj.strftime("%Y-%m-%d")
```

```
        notes_list.append(f"- {title} (created: {formatted_date})")

    return "Available notes:\n" + "\n".join(notes_list)
```

This function lists all the available notes from the database. If no notes are available in the database, it returns a "no notes found" message.

Update Note

Here's the code for updating a note:

```
def update_note(title, content):
    """Update an existing note"""
    conn = sqlite3.connect(DB_PATH)
    cursor = conn.cursor()

    # Check if the note exists
    cursor.execute("SELECT id FROM notes WHERE title = ?",
    (title,))
    if not cursor.fetchone():
        conn.close()
        return f"No note found with title '{title}'."

    # Update the note
    current_time = datetime.now().isoformat()
    cursor.execute(
        "UPDATE notes SET content = ?, updated_at = ? WHERE
        title = ?",
        (content, current_time, title)
    )

    conn.commit()
    conn.close()

    return f"Note '{title}' updated successfully."
```

The function above updates a note if it exists.

Delete Note

Here's the code for deleting a note:

```python
def delete_note(title):
    """Delete a note by its title"""
    conn = sqlite3.connect(DB_PATH)
    cursor = conn.cursor()

    # Check if the note exists
    cursor.execute("SELECT id FROM notes WHERE title = ?",
    (title,))
    if not cursor.fetchone():
        conn.close()
        return f"No note found with title '{title}'."

    # Delete the note
    cursor.execute("DELETE FROM notes WHERE title = ?",
    (title,))

    conn.commit()
    conn.close()

    return f"Note '{title}' deleted successfully."
```

The function searches for a note with the title you specify and deletes the note, if available.

Search Note

Here's the code for searching a note:

```python
def search_notes(query):
    """Search for notes containing the query in title or
    content"""
```

```python
    conn = sqlite3.connect(DB_PATH)
    cursor = conn.cursor()

    search_pattern = f"%{query}%"
    cursor.execute(
        "SELECT title, created_at FROM notes WHERE title LIKE ? "
        "OR content LIKE ? ORDER BY created_at DESC",
        (search_pattern, search_pattern)
    )
    notes = cursor.fetchall()

    conn.close()

    if not notes:
        return f"No notes found matching '{query}'."

    notes_list = []
    for title, created_at in notes:
        # Format the date to be more readable
        date_obj = datetime.fromisoformat(created_at)
        formatted_date = date_obj.strftime("%Y-%m-%d")
        notes_list.append(f"- {title} (created: {formatted_
        date})")

    return f"Notes matching '{query}':\n" + "\n".
    join(notes_list)
```

This function will fetch the notes using the title or the content.

```python
print(save_note("Test note", "This is a test note"))
print(get_note("Test note"))
print(list_notes())
print(update_note("Test note", "Updated test note"))
print(search_notes("test"))
print(delete_note("Test note"))
```

CHAPTER 7   SAVING NOTES WITH MELISSA

To test the plugin, execute the notesManager.py file.
python3 src/plugins/notesManager.py

You should get similar output to this:

```
Note 'Test note' saved successfully.
Note: Test note
Created: 2025-06-29T15:44:05.380899

This is a test note
Available notes:
- Test note (created: 2025-06-29)
Note 'Test note' updated successfully.
Notes matching 'test':
- Test note (created: 2025-06-29)
Note 'Test note' deleted successfully.
```

You have successfully created a plugin that allows you to manage your notes. Before integrating it with Melissa, make sure to remove the print statements added at the bottom of the file.

Integrating the Notes Plugin with Melissa

You have created the notes plugin; you now need to integrate it with Melissa. This involves updating the `logic.py` file to include the new plugin in the available functions and tools list.

```
from plugins.notesManager import save_note, get_note, list_notes, update_note, delete_note, search_notes

def logic(messages):
    """
    Process messages and handle tool calls
    Returns appropriate response from the LLM.
    """
```

```python
available_functions = {
    'get_date_time': get_date_time,
    'get_hackernews_info': get_hackernews_info,
    'get_weather': get_weather,
    'save_note': save_note,
    'get_note': get_note,
    'list_notes': list_notes,
    'update_note': update_note,
    'delete_note': delete_note,
    'search_notes': search_notes
}

# Define tool schema more explicitly
tools = [
    # ... existing tools ...
    {
    "type": "function",
        "function": {
            "name": "save_note",
            "description": "Save a note with a title and content",
            "parameters": {
                "type": "object",
                "properties": {
                    "title": {
                        "type": "string",
                        "description": "The title of the note"
                    },
                    "content": {
                        "type": "string",
```

```
                                            "description": "The
                                            content of the note"
                                        }
                                    },
                                    "required": ["title", "content"]
                                }
                            }
                        },
                        {
                            "type": "function",
                            "function": {
                                "name": "get_note",
                                "description": "Retrieve a note by
                                its title",
                                "parameters": {
                                    "type": "object",
                                    "properties": {
                                        "title": {
                                            "type": "string",
                                            "description": "The
                                            title of the note to
                                            retrieve"
                                        }
                                    },
                                    "required": ["title"]
                                }
                            }
                        },
                        {
                            "type": "function",
                            "function": {
```

```
                    "name": "search_notes",
                    "description": "Search for notes containing
                    a specific term",
                    "parameters": {
                            "type": "object",
                            "properties": {
                                    "query": {
                                        "type": "string",
                                        "description": "The
                                        search term to find
                                        in notes"
                                    }
                            },
                        "required": ["query"]
                    }
                }
        },
            {
            "type": "function",
                    "function": {
                        "name": "list_notes",
                        "description": "List all available notes",
                        "parameters": {
                                "type": "object",
                                "properties": {},
                            "required": []
                        }
                }
        }, {
            "type": "function",
                    "function": {
```

```
                        "name": "update_note",
                        "description": "Update an existing note
                        with new content",
                        "parameters": {
                                "type": "object",
                                "properties": {
                                        "title": {
                                                "type": "string",
                                                "description": "The
                                                title of the note
                                                to update"
                                        },
                                        "content": {
                                                "type": "string",
                                                "description": "The new
                                                content for the note"
                                        }
                                },
                                "required": ["title", "content"]
                        }
                }
        }, {
                "type": "function",
                "function": {
                        "name": "delete_note",
                        "description": "Delete a note by
                        its title",
                        "parameters": {
                                "type": "object",
                                "properties": {
                                        "title": {
```

CHAPTER 7   SAVING NOTES WITH MELISSA

```
                            "type": "string",
                            "description": "The
                            title of the note
                            to delete"
                        }
                    },
                    "required": ["title"]
                }
            }
        }
    ]
    # ... rest of the function ...
```

Next, update the system prompt in the main.py file to include information about the new notes plugin:

```
messages = [
    {
        'role': 'system',
        'content':
        '''
        You are an AI assistant designed to be helpful and
        efficient. You have tools to assist with specific tasks.

        **TOOL USAGE GUIDELINES:**
        * If a user asks a question that your tool is designed
          to answer, you MUST use the tool.
        * You MUST use your available tools whenever a user's
          request directly and clearly maps to a tool's
          capability.
```

CHAPTER 7 SAVING NOTES WITH MELISSA

AVAILABLE TOOLS:

1. **get_current_date:** Get the current date.
***Instruction:** You MUST use this tool for ANY request about the current date (e.g., "What's today's date?"). Do not attempt to answer date questions from your own knowledge.

2. **get_current_time:** Get the current time.
***Instruction:** You MUST always use this tool for ANY request about the current time (e.g., "What time is it?"). Do not attempt to answer time questions from your own knowledge or conversation history.

3. **get_hackernews_info:** Get top stories from HackerNews.
***Instruction:** You MUST use this tool for ANY request about the top stories on HackerNews (e.g., "What are the top stories on HackerNews?"). Do not attempt to answer HackerNews questions from your own knowledge.

4. **get_weather:** Get weather information for a specific city.
***Instruction:** You MUST use this tool for ANY request about the weather in a specific city (e.g., "What's the weather in London?"). Do not attempt to answer weather questions from your own knowledge.

5. **save_note:** Save a new note.
***Instruction:** You MUST use this tool for ANY request to save a new note (e.g., "Save a note about my meeting tomorrow"). Do not attempt to answer notes questions from your own knowledge.

6. **get_note:** Retrieve a note by its title.
***Instruction:** You MUST use this tool for ANY request to retrieve a specific note (e.g., "Show me my note about the meeting"). Do not attempt to answer notes questions from your own knowledge.

7. **list_notes:** List all available notes.
***Instruction:** You MUST use this tool for ANY request to list all notes (e.g., "What notes do I have?"). Do not attempt to answer notes questions from your own knowledge.

8. **update_note:** Update an existing note.
***Instruction:** You MUST use this tool for ANY request to update an existing note (e.g., "Update my note about the meeting"). Do not attempt to answer notes questions from your own knowledge.

9. **delete_note:** Delete a note by its title.
***Instruction:** You MUST use this tool for ANY request to delete a note (e.g., "Delete my note about the meeting"). Do not attempt to answer notes questions from your own knowledge.

10. **search_notes:** Search for notes by a keyword.
***Instruction:** You MUST use this tool for ANY request to search notes (e.g., "Search for notes about John"). Do not attempt to answer notes questions from your own knowledge.

For all other questions not covered by your tools, respond naturally.

CHAPTER 7 SAVING NOTES WITH MELISSA

EXAMPLES:

User: "What time is it?"
Assistant: The current time is 12:34

User: "Tell me today's date"
Assistant: Today's date is 17 Dec, 2025

User: "What are the top stories on HackerNews?"
Assistant: Here are the top stories from HackerNews:
   1. [Story Title]
   Author: [Author Name]
   2. [Story Title]
   Author: [Author Name]

User: "What's the weather in London?"
Assistant: Here's the current weather in London:
   Conditions: Partly cloudy
   Temperature: 12 C
   Humidity: 65%
   Min Temp: 4 C
   Max Temp: 14 C
   Feels Like: 11 C

User: "Save note Meeting Tomorrow: I have a meeting with John at 2pm to discuss the project"
Assistant: Note 'Meeting Tomorrow' saved successfully.

User: "What notes do I have?"
Assistant: Here are your available notes:
   - Meeting Tomorrow (created: 2023-12-17)
   - Shopping List (created: 2023-12-16)

User: "Show me my note about the meeting"
Assistant: Note: Meeting Tomorrow
    Created: 2023-12-17T14:30:45.123456
    I have a meeting with John at 2pm to discuss
    the project

User: "Search for notes about John"
Assistant: Notes matching 'John':
    - Meeting Tomorrow (created: 2023-12-17)
    - Project Ideas (created: 2023-12-15)

User: "Update my note about the meeting. I will meet John at 3pm instead of 2pm"
Assistant: Note 'Meeting Tomorrow' updated successfully.

User: "Delete my note about the meeting"
Assistant: Note 'Meeting Tomorrow' deleted successfully.

User: "Hello, my name is John"
Assistant: Hello John, how can I help you today?
    '''
    },
]

Testing the Notes Plugin

You have created the notes plugin and provided it as a tool to Melissa. To test it, start the Ollama server and execute the src/main.py file.
ollama serve
python3 src/main.py

CHAPTER 7   SAVING NOTES WITH MELISSA

You can now interact with Melissa and use the notes plugin to save, retrieve, list, update, delete, and search notes. Here are some examples of commands you can use:

> "Save note Meeting Tomorrow: I have a meeting with John at 2pm to discuss the project"

> "Get note Meeting Tomorrow"

> "List notes"

> "Update note Meeting Tomorrow: I have a meeting with John at 3pm to discuss the project"

> "Delete note Meeting Tomorrow"

> "Search notes John"

You are now able to manage notes with Melissa. If you noticed the project structure, new directory `data` will be added automatically. This is the directory that stores the `.sqlite` file. Add this directory to the `.gitignore` file. This will prevent you from pushing this directory to GitHub and thus keep your notes secure.

EXERCISE

The notes plugin is working well. However, if you ask it to save the Hacker News story for later reading, the plugin can't do that yet. Can you update the plugins to enable this functionality?

Solution

To allow the notes plugin to save the URL of the Hacker News story in the notes, you need to first update the Hacker News plugin. This plugin returns only the title and author. Update the plugin to return the URL as well.

CHAPTER 7   SAVING NOTES WITH MELISSA

The get_hackernews_info method in the getHackerNews.py file should be updated as follows:

```python
def get_hackernews_info(query: str) -> str:
    """Process user query and return HackerNews information"""
    # Default to 5 stories if no number specified
    limit = 5

    # Check if user specified a number of stories
    words = query.lower().split()
    try:
        if 'top' in words and len(words) > words.index('top') + 1:
            limit = int(words[words.index('top') + 1])
            limit = min(limit, 20)  # Cap at 20 stories
    except ValueError:
        pass

    stories = fetch_top_stories(limit)
    if not stories:
        return "Sorry, I couldn't fetch stories from HackerNews"

    response = f"Top {len(stories)} stories from HackerNews:\n\n"
    for i, story in enumerate(stories, 1):
        title = story.get('title', 'No title')
        author = story.get('by', 'unknown')
        url = story.get('url', '')

        response += f"{i}. {title}\n"
        response += f"   Author: {author}\n"
        if url:
            response += f"   URL: {url}\n"

    return response
```

Now that the plugin returns the URL, you can save it as a note.

To see this in action, start the program. Ask Melissa to return the top five stories from Hacker News and then ask the assistant to save any of the returned stores to the notes. Melissa will add the story to your notes, which you can catch up on later!

Get the Code

Just like the previous chapters, if you are running into issues or want to just get the code, you can find it on the GitHub repository: `https://github.com/Melissa-AI/melissa-v2`. Clone the repository, and check out to the `chapter-7` branch. You can use the code as is. Just make sure to create and activate a virtual environment and install the packages. You will also need the API key from OpenWeatherMap.

Summary

In this chapter, you learned about local data storage and created a plugin that allows Melissa to save, retrieve, list, update, delete, and search notes. You used SQLite as the database system to store and manage the notes efficiently.

With this plugin, Melissa can now help you keep track of important information, ideas, or reminders without leaving your conversation. You can save URLs of interesting articles, jot down ideas for future projects, or create to-do lists, all through your conversation with Melissa.

You have built quite a few plugins so far and have learned all the necessary basics. In the next chapter, I will share with you a few ideas for plugins that you can create for yourself. I have also shared blueprints on how to build those plugins.

CHAPTER 8

What More Can You Build?

Melissa, your virtual assistant can now help you answer questions, tell you the current date and time, give you weather updates, fetch the top stories from Hacker News, and manage your notes!

In the previous chapters, you learned about the plugin architecture and created plugins for specific tasks. In this chapter, you will learn about some other plugins that you can build. With the plugin idea, you will also understand the business logic. This will help you create the plugins yourself.

Remind Yourself to Call Dad

Reminders are a fantastic way to make sure you aren't forgetting that chore you have been putting off for a while. Melissa can help you get your to-do list done, whether it is calling your dad, or sending out that important email.

There are different approaches to building the reminder plugin. The simplest way is to run a cron scheduler. A cron scheduler can execute a program at defined time intervals. You can define the scheduler to run every second/minute/hour/day or even at specific time of the day or a specific day of the week.

The flow of the plugin can be as follow:

> Create a scheduler (e.g., run every day at 8 a.m.)
>
> Check the to-do list and filter it to show only the tasks that are due for the day
>
> Send the filtered to-dos to the LLM
>
> Melissa speaks the response to you

Once created, this plugin can be highly useful. Add a to-do of "Call dad on 15th July" and Melissa will remind you about this on the 15th of July.

What's Next on My Schedule?

If you manage your life with a calendar, a plugin that connects to your calendar can be a life-saver. You can ask Melissa about all the events of the day or ask to add a new calendar invitation for a meeting.

To build this plugin, you need to provide Melissa with access to your calendar. It is important to note that you should take proper security measures when providing access to the resources that Melissa might need.

Using the calendar management plugin, your virtual assistant can then help you manage your schedule!

Control the Lights with Voice

While you can ask Melissa to help you manage your schedule and to-do list, you can also use it to control lights and other home appliances. Before going to bed, ask Melissa to turn off all the lights or switch them to sleep mode, and Melissa will take care of it.

Apart from just controlling your lights, there is a wide range of home automations that you can ask Melissa to handle. Another example is setting the brightness and even the color of the light based on the outside brightness. Provide Melissa with the details of how bright it is outside and let Melissa set the lights in your home to perfect brightness, temperature, and color!

Based on the types of appliances you are using, there are different ways to achieve this. One of the most popular ways is to use Home Assistant (https://www.home-assistant.io/). Home Assistant is an open-source project that allows you to control your appliances and supports thousands of devices. Configure Home Assistant and add your supported devices. Provide Melissa access to Home Assistant via the API, and you should be able to control the appliances with your voice!

When Is the Next Bus?

If you live in a city that has a good public transport network, this one is for you! You can keep track of the train/tram/bus that you need to catch. Simply ask Melissa when the bus is scheduled to arrive, and Melissa will tell you.

If your city provides you with a public API to fetch such details, you can use that and build a plugin. However, if the city you live in doesn't provide a public API, you can use the Google Maps API (https://developers.google.com/maps/documentation/routes/transit-route) to fetch these details.

With this plugin, you can always check when the next bus/train/tram is arriving before you leave for the station.

Share Your Thoughts with the World!

If you enjoy using social media platforms like X (formerly Twitter) and Bluesky, you can create a plugin that tweets for you! This plugin can also help with fetching the latest posts on your feed and reading them out loud.

CHAPTER 8   WHAT MORE CAN YOU BUILD?

Let's understand how you would create such a plugin. Both X and Bluesky have different approaches. In this section, you will learn how you can create plugins for both the platforms.

For X, you can use the popular library Tweepy (https://www.tweepy.org/). Tweepy provides you with all the functions that you require to create this plugin. You will also need to create OAuth credentials from the X Developer platform. Make sure to securely store these credentials. Once you have these configured, you can use the home_timeline method (https://docs.tweepy.org/en/stable/api.html#get-tweet-timelines) to fetch the tweets and the update_status method (https://docs.tweepy.org/en/stable/api.html#tweepy.API.update_status) to post a tweet.

Bluesky uses the AT Protocol to be a truly decentralized social media platform. Hence, for Bluesky, you can use the AT Protocol Python library atproto (https://atproto.blue/en/latest/). After you install the library, you will need to set it up using your Bluesky API username and password. You don't have to use your original Bluesky password, you can create a new password for this plugin, which is separate from your original password. Once you have configured the atproto client, you can use the send_post method (https://atproto.blue/en/latest/atproto_client/client.html#atproto_client.client.client.Client.send_post) to create a new post and the get_posts method (https://atproto.blue/en/latest/atproto_client/client.html#atproto_client.client.client.Client.get_posts) to fetch posts from your feed.

You can do a lot more! Based on your preferences, you can use different methods provided by the libraries to interact with the social media platforms.

CHAPTER 8   WHAT MORE CAN YOU BUILD?

Play My Favorite Song

While you have been building plugins that make you productive, you can also build plugins to have some fun! One such idea is to use Melissa to control your Spotify player. Spotify has an API that allows you to manage the Spotify Player (https://developer.spotify.com/documentation/web-api/reference/start-a-users-playback). You can use the API to ask Melissa to play/pause a song, play the next song, or play a specific song or a playlist. While the API allows you to control the player, it doesn't allow you to play the songs directly from your program. You will need the Spotify player running on your phone, computer, or any other supported device.

Similarly, if you're running a Plex Media Server to organize and stream your personal music, movie, and TV show collections, you can extend Melissa to control your Plex playback. Plex offers a robust API that allows for detailed interaction with your server and connected Plex clients.

To build a Plex plugin for Melissa, you would typically do the following:

> **Identify your Plex server and client**: You'll need the address of your Plex server and the name or ID of the Plex client (e.g., "Living Room TV," "John's Phone," "Plex Web Chrome") you want to control.
>
> **Authentication**: The Plex API requests require an authentication token. You'll need to obtain this token from your Plex server.
>
> **Utilize the Plex API (or a library)**: While you can make direct HTTP requests to the Plex API, a more convenient approach is to use a Python library like python-plexapi (available on PyPI: `pip install PlexAPI`). This library simplifies connecting to your server, finding media, selecting clients, and controlling playback.

Implement commands: With access to the API, you can create Melissa commands to:

- Play, pause, or stop playback on a specific Plex client
- Skip to the next or previous track/episode
- Seek to a specific point in the media
- Adjust the volume
- Play specific songs, albums, artists, or playlists from your music library
- Start movies or specific TV show episodes

Ask "What's playing?" to get information about the current media.

This allows you to have voice control over your self-hosted media library, making it even more accessible and enjoyable. Just like with Spotify, the actual media playback happens on the chosen Plex client device, while Melissa acts as the voice-activated remote control.

You can take this a step further! You can play the music on your Raspberry Pi, where you are running Melissa. You can use the same Raspberry Pi as a Plex Media client.

Summary

With the plugin architecture, you can make Melissa more capable of performing tasks on your behalf. The virtual assistant can help you be more productive. Use the APIs of the services that you use every day and create a plugin to provide Melissa with the capability.

CHAPTER 8 WHAT MORE CAN YOU BUILD?

You have already learned how to use APIs and created plugins that allow Melissa to interact with the platform. You can use that knowledge and build plugins that can help you be more productive.

In the next chapter, you will learn how to take your virtual assistant and set it up on your Raspberry Pi (RPi). You will learn to set up your RPi from scratch and integrate Melissa to use it from the RPi.

CHAPTER 9

Integrating the Software with Raspberry Pi, and Next Steps

In this chapter, you will learn to get a Raspberry Pi 5 running and integrate your software, Melissa, to work within its operating system, Raspberry Pi OS. You will see how this proof-of-concept software can be scaled to make a full-fledged assistant, and you will go through various enhancements and use cases for Melissa.

Up to this point, you have successfully developed a virtual assistant that listens to you, understands what you say to some extent, and speaks back to you. It can also do a lot of useful things for you, such as tell you the time and weather, save notes, and so on. You have Melissa running successfully on your desktop or laptop. Now it's time to set up Melissa on a Raspberry Pi 5 so that it can contribute to making the Internet of Things (IoT) smarter.

First, you need to set up your Raspberry Pi (RPi). Even if you don't have an RPi 5 yet, you should still go through this chapter; it will broaden your views on how you can scale Melissa to make it more useful and how you can employ Melissa in different scenarios to make your devices smarter.

CHAPTER 9   INTEGRATING THE SOFTWARE WITH RASPBERRY PI, AND NEXT STEPS

Setting Up a Raspberry Pi 5

The Raspberry Pi 5 comes with the essential tools, making it an excellent choice for running Melissa. While the core concept remains the same, the setup process and some requirements have been updated.

As you've likely noticed, the Raspberry Pi 5 still comes as a bare board. You'll need a few essential accessories:

> **Power supply**: The Raspberry Pi 5 requires a high-quality 5V/5A USB-C Power Delivery (PD) power supply. Using an older or lower-amperage power supply may lead to instability or prevent the Pi 5 from booting correctly. The official Raspberry Pi 27W USB-C Power Supply is recommended.
>
> **MicroSD card**: A reliable microSD card (at least 16GB, 32GB recommended, Application Class 2 (A2) rated for better performance) is needed to install the operating system.
>
> **MicroSD card reader**: This is needed to write the operating system to the card using your computer.
>
> **Connectivity**: Use an Ethernet cable for a wired network connection (recommended for initial setup and reliability) or use the built-in Wi-Fi.

These are the peripherals that are optional but recommended:

> **USB keyboard and mouse**: You can connect the USB keyboard with your Raspberry Pi device and use it to control the device.
>
> **Display with an HDMI input**: Note that the Raspberry Pi 5 has two micro-HDMI ports, so you might need a micro-HDMI to standard HDMI cable or adapter.

CHAPTER 9   INTEGRATING THE SOFTWARE WITH RASPBERRY PI, AND NEXT STEPS

Active cooling: Due to its increased power, the Raspberry Pi 5 typically requires active cooling. The official Raspberry Pi Active Cooler or a case with an integrated fan is highly recommended to prevent overheating and ensure optimal performance.

Case: This will protect the board.

Installing the Operating System (Raspberry Pi OS)

The recommended way to install Raspberry Pi OS (the official operating system, based on Debian) is using the Raspberry Pi Imager tool. This simplifies the process significantly compared to older methods.

Download and install Raspberry Pi Imager. Get the latest version for your computer (Windows, macOS, or Linux) from the official Raspberry Pi website (https://www.raspberrypi.com/software/).

Next, insert the microSD card into your computer's reader and open Raspberry Pi Imager (see Figure 9-1).

CHAPTER 9   INTEGRATING THE SOFTWARE WITH RASPBERRY PI, AND NEXT STEPS

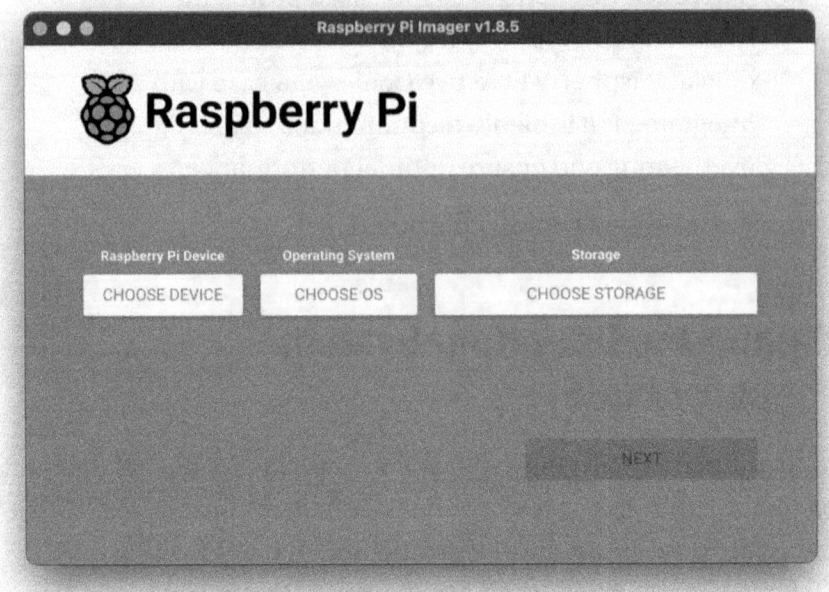

Figure 9-1. *Raspberry Pi Imager*

Click CHOOSE DEVICE and select Raspberry Pi 5, as shown in Figure 9-2.

CHAPTER 9   INTEGRATING THE SOFTWARE WITH RASPBERRY PI, AND NEXT STEPS

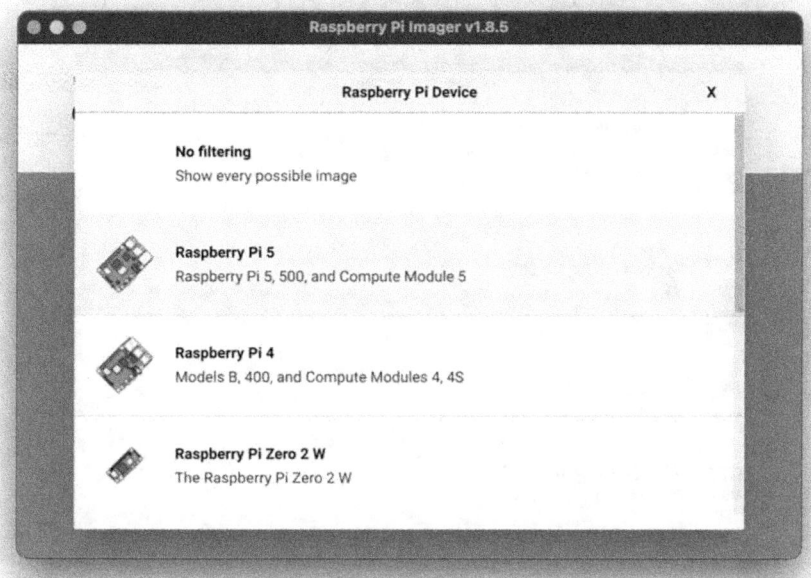

Figure 9-2. *Selecting the Raspberry Pi device*

Click CHOOSE OS and then choose the recommended version, typically Raspberry Pi OS (64-bit) or Raspberry Pi OS Full (64-bit) for a desktop experience, as shown in Figure 9-3.

CHAPTER 9   INTEGRATING THE SOFTWARE WITH RASPBERRY PI, AND NEXT STEPS

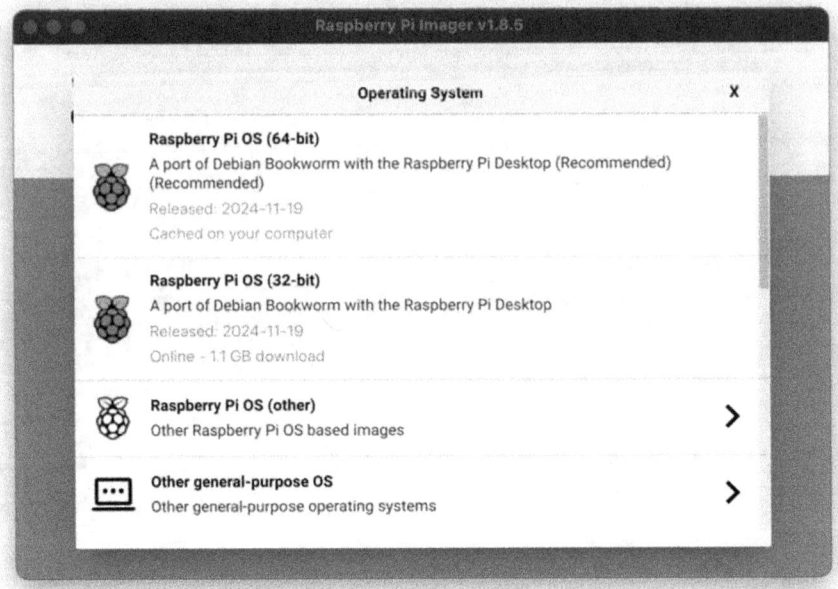

Figure 9-3. *Selecting Raspberry Pi OS*

Click CHOOSE STORAGE and select your microSD card. Warning: All data on the selected card will be erased.

Preconfigure (Highly Recommended for Headless Setup)

Click NEXT. When prompted to apply OS customization settings, select EDIT SETTINGS. Customize these settings, as shown in Figure 9-3:

- **Set hostname**: Give your Pi a network name (e.g., melissa-pi).

- **Set username and password**: Create a user account and a strong password. Avoid using the old default "pi" username for security reasons. Remember these credentials!

CHAPTER 9   INTEGRATING THE SOFTWARE WITH RASPBERRY PI, AND NEXT STEPS

- **Configure wireless LAN**: Enter your Wi-Fi network name (SSID) and password if you plan to use Wi-Fi. Set the correct WLAN country.

Figure 9-4. Customizing settings

On the Services tab, do the following:

- Enable SSH: Check this box and choose "Use password authentication" for simplicity initially. This allows remote terminal access.
- Click SAVE.

To write the image, do the following:

- Click YES to confirm the settings.
- Click YES again to confirm you want to erase the microSD card.

The Imager will now write the OS to the card and verify it. This may take several minutes.

Once the Imager finishes, safely eject the microSD card from your computer. Insert the card into the microSD slot on your Raspberry Pi 5.

Connect your keyboard, mouse, and display (using a micro-HDMI port) if you are not running headless. Connect the Ethernet cable if using a wired connection. Ensure the active cooler/fan is properly connected if using one.

Connect the 5V/5A USB-C power supply to power on the Pi. The Pi 5 also has a dedicated power button near the USB-C port you can use.

The Pi will boot up. The first boot might take slightly longer. If you connect to a display, follow any on-screen setup prompts (like confirming language/time zone, connecting to Wi-Fi if not pre-configured, and checking for software updates).

Connecting Remotely (Headless Setup)

If you set up your Pi without a display (headless) and enabled SSH via the Imager, you can now connect to it from another computer on the same network.

CHAPTER 9   INTEGRATING THE SOFTWARE WITH RASPBERRY PI, AND NEXT STEPS

Find the Pi's IP address to connect to it. Your router's administration page usually lists connected devices and their IP addresses (look for the hostname you set, e.g., melissa-pi). Alternatively, you can use network scanning tools.

You can also connect to the Pi using the hostname you set earlier.

From Linux or macOS, open a terminal and type the following:

```
$ ssh your_username@hostname.local
```
(e.g., ssh assistant@melissa-pi.local)

Or type the following:

```
$ ssh your_username@YOUR_PI_IP_ADDRESS
```
(e.g., ssh assistant@192.168.1.15)

From Windows, use PowerShell or the command prompt and type the same ssh command as shown earlier. Alternatively, use an SSH client like PuTTY.

The first time you connect, you'll be asked to confirm the host's authenticity; type yes.

Enter the password you set during the Imager configuration.

You should now be logged into your Raspberry Pi's command line.

Setting Up VNC (Remote Desktop Access)

If you want graphical desktop access remotely, you can use VNC. RealVNC is included with Raspberry Pi OS and is easy to enable.

Connect to your Pi via SSH or open a terminal directly on the Pi if using a display.

CHAPTER 9   INTEGRATING THE SOFTWARE WITH RASPBERRY PI, AND NEXT STEPS

Run the following command:

```
$ sudo raspi-config
```

Navigate using the arrow keys to Interface Options ➤ VNC ➤ Yes to enable the VNC server, as shown in Figure 9-5 and Figure 9-6.

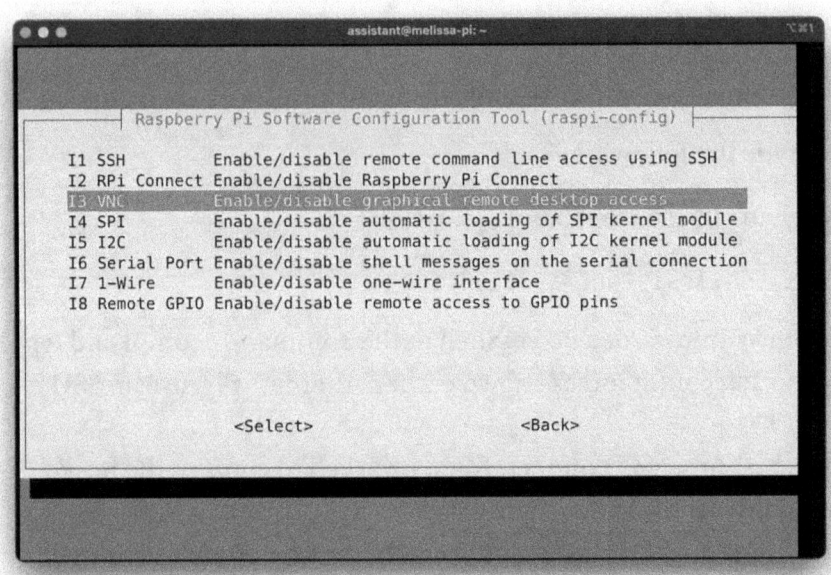

Figure 9-5. *raspi-config*

CHAPTER 9   INTEGRATING THE SOFTWARE WITH RASPBERRY PI, AND NEXT STEPS

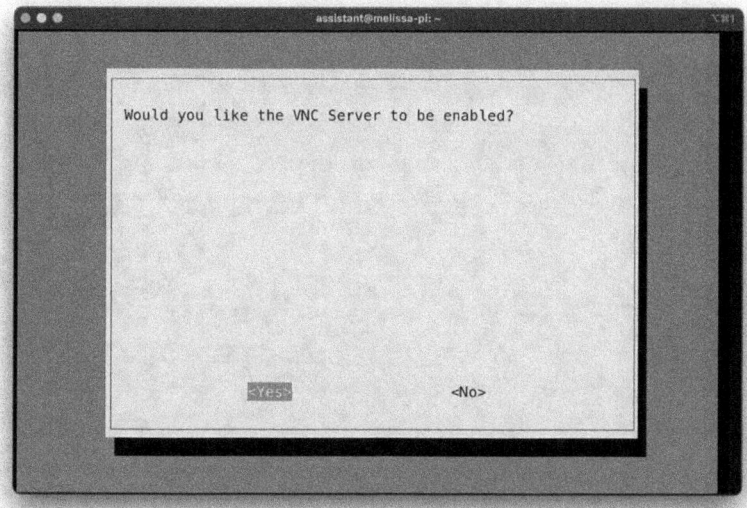

Figure 9-6. *VCN enable confirmation*

Select OK and then Finish. Reboot if prompted (sudo reboot).

Note   With newer Raspberry Pi OS versions (Bookworm and newer), you might need to ensure the system is using the X11 display server instead of Wayland for RealVNC compatibility, especially for headless setups. In raspi-config, go to Advanced Options and then Wayland and select X11 if you encounter issues.

On the computer from which you want to control the Pi, download and install RealVNC Viewer (https://www.realvnc.com/en/connect/download/viewer/).

Open VNC Viewer on your computer and from the menu bar select File ➤ New connection. Enter the IP address of your Pi in the VNC Server field. Enter an easy to identify name in the Name field and click OK, as shown in Figure 9-7.

121

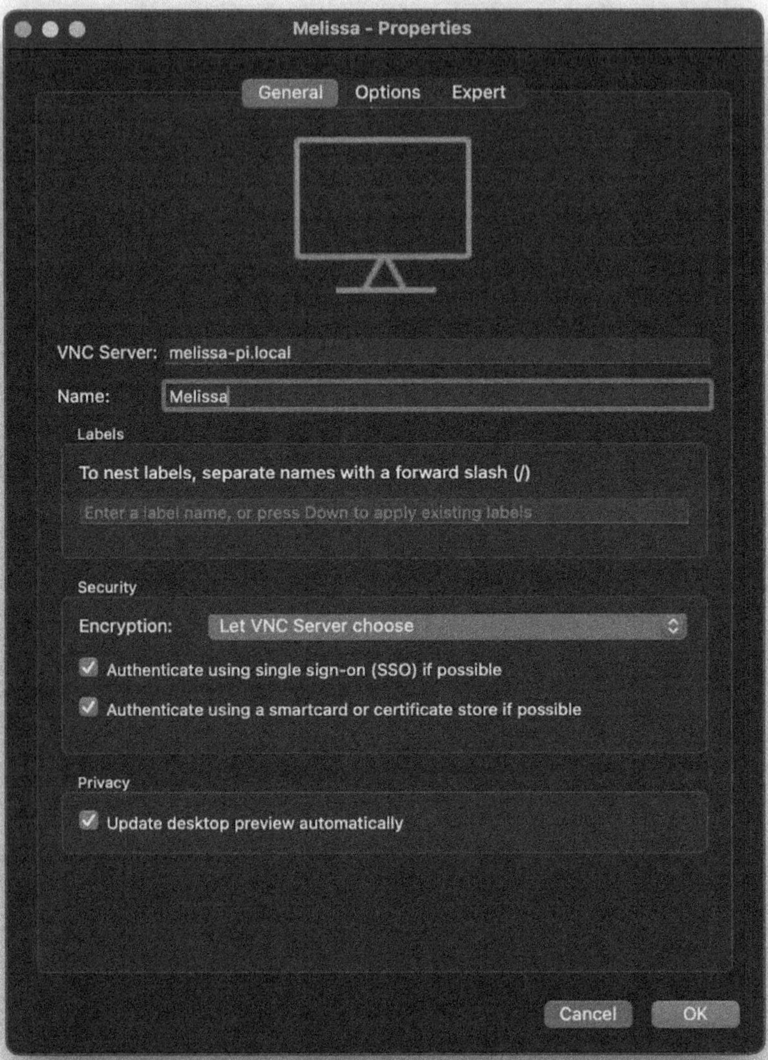

CHAPTER 9   INTEGRATING THE SOFTWARE WITH RASPBERRY PI, AND NEXT STEPS

Figure 9-7.   Connecting to Raspberry Pi via VNC

CHAPTER 9   INTEGRATING THE SOFTWARE WITH RASPBERRY PI, AND NEXT STEPS

You can find the newly added connection in the RealVNC Viewer application. Double-click the connection, and when prompted, enter the username and password you configured earlier.

Adding New Components to the Raspberry Pi 5

To effectively interact with Melissa running on your Raspberry Pi 5, you'll need some additional components:

> **Case**: As mentioned, a case is crucial for protecting your Pi 5, especially given the recommendation for active cooling, which often comes integrated into cases designed for the Pi 5.
>
> **Microphone**: To give voice commands to Melissa, a USB microphone is needed. Most standard USB microphones should work with the Raspberry Pi 5, but check the compatibility if unsure. Connect it to one of the available USB ports (the Pi 5 has two USB 3.0 and two USB 2.0 ports).
>
> **Speakers/headphones**: To hear Melissa's responses, you'll need speakers or headphones. The Raspberry Pi 5 does not have the 3.5mm audio jack found on older models like the Pi 4. You will need to use the following:
>
>> **USB speakers/headphones**: Connect speakers or a headset via one of the USB ports.
>>
>> **HDMI audio**: If your connected display has built-in speakers, audio can be output via the micro-HDMI connection.

Bluetooth speakers/headphones: Pair a Bluetooth audio device using the Pi's built-in Bluetooth 5.0.

DAC HAT: Use a digital-to-analog converter hardware attached on top (HAT) board if you require a high-quality analog audio output via the GPIO pins.

Ensure you select the correct audio input (your USB microphone) and output (USB/HDMI/Bluetooth/HAT) device within the Raspberry Pi OS audio settings (usually accessible via the volume icon in the taskbar when using the desktop environment).

Setting Up Melissa

Now that your Raspberry Pi 5 is set up and you can access it (either directly or remotely), you can install Melissa.

To set up Melissa, you first have to repeat all the steps such as installing third-party utilities such as PortAudio, PyAudio, Ollama, espeak, and mpg123. You can find the complete list of things you need to install in the Wiki of Melissa repository: https://github.com/Melissa-AI/melissa-v2/wiki. Note: Ensure these dependencies are compatible with the 64-bit Raspberry Pi OS and update the installation commands if necessary.

```
$ sudo apt-get install python3-pyaudio portaudio19-dev
  python3-all-dev espeak
$ curl -fsSL https://ollama.com/install.sh | sh
$ ollama pull qwen3:1.7b
```

Now you need to transfer your code repository from your local development environment to the RPi 5. I recommend you fork the official repository linked in the previous paragraph and clone it with the help of Git.

CHAPTER 9   INTEGRATING THE SOFTWARE WITH RASPBERRY PI, AND NEXT STEPS

Enter the following command on your RPi 5 terminal to clone Melissa:

```
$ git clone https://github.com/Melissa-AI/melissa-v2.git
```

There is another advantage to setting up Melissa by cloning the repository. You can install all the required Python modules listed in `requirements.txt` by entering the following command (navigate into the `Melissa-Core` directory first):

```
$ pip3 install -r requirements.txt
```

If you are transferring Melissa from your local environment manually instead of cloning, you must export a list of the Python modules you have installed via `pip`. You can export all this information to a text file by entering the following command on your local terminal:

```
$ pip3 freeze > requirements.txt
```

Then transfer this `requirements.txt` file along with your code to the RPi 5 and run the following:

```
$ pip3 install -r requirements.txt
```

You also need to set up the configuration file. Navigate into the Melissa directory on your RPi 5 and update the `profile.json` file with your details. You also need to add the API credentials. Rename the `.env.copy` file to `.env` and add the API credentials.

```
$ cp .env.copy .env
```

Once you have successfully set up your development environment, open `profile.json` to customize the file and add details about yourself.

You should now be able to run Melissa on Raspberry Pi OS. Make sure you start the Ollama server before you run the program. If you get any error messages, you may be missing a component or encountering a compatibility issue with the newer OS or Python versions. Try to debug

125

Melissa using the error messages provided by Python's interpreter. Refer to the Melissa project's documentation or issue tracker for potential solutions specific to Raspberry Pi 5 or Raspberry Pi OS Bookworm.

Making Melissa Better Each Day!

This section talks about how you can add new and more complex functionalities to make Melissa better each day. The preferred way to do this is to contribute to the project. If you have an idea of a new plugin or ways to improve the functionality, open an issue for discussion, and submit a pull request with the changes you want. After testing your feature, we can merge it into the official codebase.

Let's go through some sample features/subprojects that you can own and that would make a big impact on Melissa's functionality:

>**Tests**: No code repository is truly professional unless it has tests built for it. Melissa currently does not have any tests that run health checks. These tests would be run by contributors before submitting their pull requests. They might include checking for errors, trailing blank spaces, PEP-8 guidelines confirmation, and much more. You could make a package of tests by creating a `test` directory in the root folder of the project.

>**Vision**: Melissa currently accepts voice input from the user, as discussed in the workflow in Chapter 1. It would be great if Melissa could gather information using cameras as well. There are large language models that can take images as input, analyze them, and based on the user input provide output. You could use such models for this purpose to add

CHAPTER 9   INTEGRATING THE SOFTWARE WITH RASPBERRY PI, AND NEXT STEPS

functionality to Melissa. You can also use OpenCV to add functionality such as detecting whether a room is empty, counting how many people are in the room, recognizing faces, converting text in photos to strings, and so on. This would redefine Melissa's current workflow.

Multidevice operation: Wouldn't it be cool to have two instances of Melissa running at the same time on different devices but communicating with each other via a server? This would require another piece of software running on a server that handles such requests for devices and connects them using keys that can be requested by a user. This is easier said than done. It would require quite a lot of programming to build the code for the cloud-based server as well as additions to Melissa-Core for handling the requests made to and from by the server code.

Native user interface: Two things that determine the success of software are its usability and looks. Currently Melissa works only via the command line. How about adding a user interface for Melissa that uses the widely used UI Framework for Python? This would help users interact with Melissa more easily. This is quite a task and would definitely require some time to construct, but it is a very high priority for the project to have a good UI.

By the time you read this, some of these functionalities may already have been constructed by contributors. However, you should still practice building these features on your own, because doing so will help you to achieve a greater understanding of the software. Feel free to discuss any new functionality that you think can make Melissa even better via GitHub issues.

Where Do I Use Melissa?

You may have the following thought: "Everything is cool, but except for R&D and on a laptop, where do I use Melissa?" Other than using Melissa on your laptop, there are a couple of sample use cases where I think Melissa can be helpful and make your devices and utilities more accessible and impressive.

- **Drones**: Many people are building drones using Raspberry Pis and drone kits that are readily available in e-commerce stores. By connecting the motors and functionality to an Arduino board and then to a RPi 5, you can control the drone's movement, direction of flight, and so on using Melissa, your voice-controlled virtual assistant. You can start the drone simply by giving voice commands and tell it to fly, land, or follow you. The possibilities are limitless with a creative mind.

- **Humanoid robots**: Humanoid robots are being developed by big corporations as well as individuals. These autonomous robots use software like Melissa to make them more interactive and, well, human like. If you plan to build a humanoid robot or any robot for that matter, you can integrate it with Melissa and build appropriate functionalities that extend Melissa to handle your robot.

CHAPTER 9   INTEGRATING THE SOFTWARE WITH RASPBERRY PI, AND NEXT STEPS

- **Burglar-detection system**: Using vision integration with OpenCV, you can detect whether someone has entered your house in your absence. Extending that functionality, you can program Melissa to take a picture of the person; alert you by sending a message that someone is in your house, along with their photo; call emergency services; and sound an alarm. Many other features can be integrated into such a system. Try brainstorming some ideas.

Summary

In this chapter, you learned how to set up a Raspberry Pi 5 and integrate Melissa into it. You also learned how to continue your learning after you finish reading this book and where you can implement this virtual assistant to make the most of your devices.

In this book, you learned about virtual assistants, other virtual assistants available on the market, how to develop a new virtual assistant, and how to make the virtual assistant speak, listen, and understand what the user says. You also learned about large language models (LLMs) and learned how to run them locally on your machine. You then built several plugins that let you talk with Melissa and ask her for information such as the weather and the time, get stories from Hacker News, and more.

I strongly encourage you to keep working on Melissa after you finish reading this book. Doing so will reinforce the concepts we have covered. Follow the principle of "Making Melissa Better Each Day!"

Index

A
Application programming interface (API), 29, 53
Artificial intelligence (AI), 28

B
BitBucket, 9
Bluesky, 106
Burglar-detection system, 129

C
ChatGPT, 28

D
Database storage, 82
Data storage
    local, 82
    SQLite, 82
Drones, 128

E
ElevenLabs, 21

F
File-based storage, 82
Function calling, 44

G
get_hackernews_info method, 101
getHackerNews.py, 54
get_weather function, 66, 72
GitLab, 9
Google assistant, 2

H
Hacker News API
    code, 62
    creating plugin, 54–57
    function calling, 53
    Melissa, integrating plugin, 58–61
    plugin architecture, 53
    public API, 54
Hallucination, 28
Humanoid robots, 128

I, J, K
Internet of Things (IoT), 111

INDEX

L

Large language models
(LLMs), 42, 129
    API, 29, 30
    computers, 27
    input-output flow, 29
    TTS engine, 28
Logic engine, 4, 26
    code, 38
    implementation, 33–37
    installing Ollama, 30–33
    LLMs, 27

M

MicroSD card, 112

N, O

Natural language processing
(NLP), 4
notesManager.py, 83
Notes plugin
    code, 102
    delete note, 88
    get note, 85, 86
    integrating with Melissa, 90,
        92–94, 96–98
    listing notes, 86
    notesManager.py, 83
    save note, 84, 85
    search note, 88, 90
    SQLite database, 84
    testing, 99
    updating note, 87

P, Q

Plugins
    APIs, 108
    build
        controlling lights with voice,
            104, 105
        Plex playback, 107, 108
        public API, 105
        reminder, 103, 104
        schedule, 104
    code, 51
    function calling, LLMs, 44–51
    LLM, 42
    project structure, 42, 43
    system APIs, 52
PyAudio, 10
Python Package Index (PyPI), 6

R

Raspberry Pi, 3
Raspberry Pi 5
    adding new components,
        123, 124
    essential accessories, 112, 113
    installing operating system,
        113–116, 118, 119
    Melissa's functionality, 126, 127
    setting up Melissa, 124, 125
    setting up VNC, 119, 121, 123

INDEX

S

Smart virtual assistants
   code, 12
   commercial, 2
   development environment,
      setting up, 5, 6
   learning methodology, 11, 12
   LLMs, 1
   Melissa, designing
      KISS principles, 7
      profile.json file, 8
      PyAudio, 10
      Github, save project, 8, 9
      virtual environment, 9, 10
      Vosk, 10
   qualities, 1
   working, 3, 4
Speech recognition engine, 4
Speech-to-text (STT), 5, 41
   components, 15
   Melissa's inception, 20
   Vosk, 16–19
Spotify, 107
SQLite, 82
stream.read() method, 20

T, U

Text-to-Speech (TTS)
      engine, 5, 41
   code, 25
   ElevenLabs, 21
   implementation, 23
   Linux systems, 22
   macOS, 21
   profile.json, 24
Tweepy, 106

V

Vosk, 10, 16

W, X, Y, Z

Weather API
   code, 79, 80
   creating weather
      plugin, 65, 66
   Melissa, temperature, 67, 69–72
   OpenWeather, 63, 64
   storing securely, API key,
      72–74, 76–78

GPSR Compliance
The European Union's (EU) General Product Safety Regulation (GPSR) is a set of rules that requires consumer products to be safe and our obligations to ensure this.

If you have any concerns about our products, you can contact us on

ProductSafety@springernature.com

In case Publisher is established outside the EU, the EU authorized representative is:

Springer Nature Customer Service Center GmbH
Europaplatz 3
69115 Heidelberg, Germany